Lecture Notes in Mathematics 1682

Editors:
A. Dold, Heidelberg
F. Takens, Groningen

Springer

Berlin
Heidelberg
New York
Barcelona
Budapest
Hong Kong
London
Milan
Paris
Santa Clara
Singapore
Tokyo

Hans-Dieter Alber

Materials with Memory

Initial-Boundary Value Problems
for Constitutive Equations
with Internal Variables

 Springer

Author

Hans-Dieter Alber
Fachbereich Mathematik
Technische Universität Darmstadt
Schlossgartenstraße 7
D-64289 Darmstadt, Germany
e-mail: alber@mathematik.tu-darmstadt.de

Cataloging-in-Publication Data applied for

Die Deutsche Bibliothek - CIP-Einheitsaufnahme

Alber, Hans-Dieter:
Materials with memory : initial boundary value problems for
constitutive equations with internal variables / Hans-Dieter Alber. -
Berlin ; Heidelberg ; New York ; Barcelona ; Budapest ; Hong Kong ;
London ; Milan ; Paris ; Santa Clara ; Singapore ; Tokyo : Springer,
1998
  (Lecture notes in mathematics ; 1682)
  ISBN 3-540-64066-5

Mathematics Subject Classification (1991):
35F25, 35L45, 35Q72, 73-02, 73E50, 73F15

ISSN 0075-8434
ISBN 3-540-64066-5 Springer-Verlag Berlin Heidelberg New York

© Springer-Verlag Berlin Heidelberg 1998
Printed in Germany

Typesetting: Camera-ready TeX output by the author
SPIN: 10649759   46/3143-543210 - Printed on acid-free paper

Dedicated to Rolf Leis
and to Jutta, Anne and Ina

Preface

Continuum mechanics offers a rising number of fascinating problems to mathematical analysis, and the interest of mathematicians in these problems has been revived and is steadily growing.

The reason behind this development is the computer. It makes more and more mathematical problems accessible to numerical treatment and thus opens an increasing number of regions to practical interest and application, which were formerly reserved to theoretical analysis. One might expect that this would diminish the interest in theoretical analysis. Fortunately, this is not the case. On the contrary, numerical studies call for a deeper analysis and thus open new fields in continuum mechanics to mathematical analysis. In a sense, by generating this renewed interest in continuum mechanics, the computer is leading mathematical analysis back to its origins in the studies of Newton, Euler and the Bernoullis.

This book also has its origins in these developments. To utilize the capacity of fast computers and of the finite element analysis in computing the deformation and stress distribution in metallic structures, constitutive equations are needed which model the properties of metals with sufficient accuracy. In mechanics and the engineering sciences, therefore, the often rather complex constitutive equations, which have been developed in the last twenty years, are of great interest. The numerical solution of the initial-boundary value problems containing these constitutive equations is now an important and central part in many engineering problems. My work in the investigation of these initial-boundary value problems with the means of mathematical analysis started, when colleagues from the engineering sciences brought these problems to my attention, and the results contained in this book have been obtained in close cooperation with them.

Of course, investigations of problems from plasticity and viscoelasticity have a long tradition in mathematical analysis, and a small number of investigations of initial-boundary value problems to these newly developed constitutive equations already exist. Generally however, in these studies the methods of the investigations and of the proofs strongly depend on the constitutive equations under consideration, and the results are often far from the results, which one might optimistically hope to obtain. The aim of this book is to improve this situation and to contribute to the development of a general mathematical theory of initial-boundary value problems to these constitutive equations.

Accordingly, the book is mainly directed to mathematicians interested in studying constitutive equations and in the mathematical analysis of the initial-boundary value problems containing such constitutive equations. However, it is hoped that readers from mechanics and the engineering sciences, who want to learn the principals on which the mathematical investigation of these initial-boundary value problems can be based, can also profit from this book. In particular, the first three chapters, parts of Chapters 5 – 8, and Chapter 9 and the Appendix should be of interest to them.

Although cooperation in the investigation of constitutive equations between specialists in mechanics, engineering and mathematics is very fruitful, it is not always without problems. One of the reasons is that in mechanics and engineering constitutive equations are judged according to their usefulness in modelling the behavior of real materials, while in mathematics constitutive equations are judged according to their mathematical properties. These different criteria used to judge and classify constitutive equations yield classes of constitutive equations, which in general do not coincide. Sometimes this is the cause of misunderstanding between specialists from different fields; mainly because these criteria are not given a-priori, they are, at least partly, research results and their importance is therefore not easily recognized by specialists from other fields. It is hoped that by displaying the mathematical side of the theory, this book will help to ease this cooperation.

However, these general remarks never applied to the cooperation with my colleagues in the Sonderforschungsbereich 298 "Deformation und Versagen bei metallischen und granularen Strukturen", where I always received very much help and support. This book could not have been written without this cooperation. Also essential was the help of many friends and co-workers. My thanks go to R. Balean, who went through the whole manuscript and corrected many errors, linguistic but also mathematical, K. Chełmiński, S. Ebenfeld, R. Farwig, M. Franzke, A. Heidrich, K. Hutter, who read part of the manuscript and gave valuable comments from the viewpoint of mechanics and thermodynamics, I. Jäpel, F. Klaus, F. Kollmann, who initiated the cooperation of scientists from engineering, mechanics and mathematics and directed my attention to these problems, O. Liess, P. Neff, I. Zahn, and to E. Schlaf and M. Tabbert, both of whom I owe special thanks for trustful cooperation over many years.

I am indebted to the support given me by the Deutsche Forschungsgemeinschaft through the Sonderforschungsbereich 298 and the Forschergruppe "Ingenieurwissenschaftliche und mathematische Analyse bruchmechanischer und inelastischer Probleme".

Darmstadt   Hans-Dieter Alber
November 1997

Contents

Chapter 1

Introduction

In this work we study systems of differential equations describing the deformation behavior of a class of materials with memory and the existence and uniqueness theory of initial-boundary value problems for these systems. The main examples of materials of the class considered are metals.

The behavior of materials with memory depends on the deformation history. Such a behavior is called inelastic. The continuum mechanical models intended to describe inelastic behavior consist of the balance laws, formulated as a system of partial differential equations, and a set of constitutive relations describing the dependence of the stress on the strain history. While the derivation of the balance laws is based on fundamental principles of physics and mechanics, there exist no such principles to allow us to derive the constitutive relations. Instead, a variety of much less well-founded principles are used to derive constitutive relations. As a result, a large number of different constitutive relations exist, valid under special assumptions and in special situations.

One of these assumptions commonly used to derive constitutive relations for metals is that a finite number of internal variables exist, whose values uniquely determine the state of the material and the deformation behavior. The constitutive relations consist of a set of differential equations for these internal variables, which determine their evolution in time. In many situations it is adequate to assume that only small deformations occur, which allows the constitutive relations to be partially linearized. In this work we study the nonlinear initial-boundary value problems resulting from these assumptions.

In mathematical analysis, these initial-boundary value problems were first treated around the year 1970 by G. Duvaut and J.L. Lions [62] and by J.J. Moreau [154, 155, 156]. Duvaut and Lions proved that the initial-boundary value problems to the constitutive relations of Prandtl-Reuss and Hencky possess unique solutions. As an essential tool, their proof uses at several instances monotonicity properties of the differential operators resulting from the use of these constitutive models. Moreau studies quasi-static initial-boundary problems to the Prandtl-Reuss law. Also he uses monotonicity properties and convex analysis to prove existence of solutions, and he observes that these problems are connected with 'moving convex sets'. Using ideas from thermodynamics, in 1975 a class of constitutive relations was defined by B. Halphen and Nguyen Quoc Son [88], which they called the *class of constitutive relations of generalized standard materials*. This class contains the models of Prandtl-Reuss and Hencky. Constitutive relations from this class naturally have some monotonicity properties, and they showed that these monotonicity properties immediately imply uniqueness of solutions for the initial-boundary value problems to such constitutive relations. No proof of existence of solutions is contained in their paper, but it was observed later that initial-boundary value problems to constitutive relations from this class can be formulated as initial value problems for evolution equations to monotone operators, where no further assumption is needed in the quasi-static case, and under a certain additional coercivity assumption in the dynamic case. Consequently, the general theory of these evolution equations can be used to prove existence and uniqueness of globally defined solutions to large initial data. An overview of these existence and uniqueness proofs and references to the literature are given in Chapter 3.

The development of computers in the last two decades and the method of finite elements make precise numerical computations possible today. In the engineering sciences, therefore, the need arose for constitutive equations more precisely describing the behavior of solids and especially metals than relatively simple models like the Prandtl-Reuss equations. This requirement led to the development of the large number of more complicated constitutive relations mentioned above. A number of examples of such constitutive relations are presented and discussed in Chapter 2.

The investigation of initial-boundary value problems to these constitutive equations poses new problems to mathematical analysis, since practically none of these newly developed constitutive relations belongs to the class of equations of generalized standard materials or yields a monotone differential operator. This is not an obvious statement, though. The statement is in accordance with the laws of thermodynamics, since monotonicity of the resulting differential operator is not a consequence of these laws; this is well known. Still, by writing the constitutive equations in a suitable form, it might be possible to obtain a monotone operator. A substantial part of this work is devoted to the demonstration that this is not the case. In fact, the following is shown: we define a class of constitutive relations, which we call of monotone type. This class includes the class of constitutive models of generalized standard materials and is a natural generalization of it, since constitutive relations from this class still possess the property that the resulting differential operator is monotone. In Chapter 8 we carefully study one of the engineering models presented in Chapter 2 and show that it is not of monotone type. As will become clear from these investigations and from our analysis of the engineering models in prior chapters, the results for this model are representative; similar results can be obtained for the other constitutive relations discussed in Section 2.

The class of constitutive relations of generalized standard materials is therefore too small, and it is necessary to study constitutive relations, which do not belong to this class. Examples for such investigations can be found in [1, 2, 36, 37, 38, 42, 39, 40, 106, 110, 111, 171]. Following modern trends in mathematical analysis, the proofs in these investigations are often based on the derivation of energy estimates and on weak convergence.

However, to us it seems natural to surmise that the theory of evolution equations for monotone operators is also a tool suited to study many of the newly developed constitutive equations. The main goal of the present work thus is to show that the applicability of this theory can be extended beyond the class of constitutive equations of generalized standard materials and even beyond the class of consitutive equations of monotone type to a larger class by what we call *transformation of interior variables*. In this way we hope to contribute to the development of a systematic theory for initial-boundary value problems, which use internal variables to describe inelastic material behavior of solids. The basic idea is to transform the given internal variables into new variables, and thereby also the system of differential equations, such that the new system is of monotone type, and allows application of the theory of evolution equations for monotone operators to prove existence and uniqueness of solutions for the transformed initial-boundary value problem, and thereby, after back transformation, also for the original problem.

Contents of this work. We start the investigations in **Chapter 2** with the precise formulation of the initial-boundary value problems of inelastic material behavior, which are studied in this work. Rate dependent and rate independent problems are defined, and in Section 2.2 several examples for constitutive equations from mechanics and the engineering sciences are presented.

In **Chapter 3** we define the classes of constitutive equations of pre-monotone type, of gradient type, of monotone type, and of monotone-gradient type. The latter class coincides with the class of constitutive equations of generalized standard materials and is a subclass of the classes of monotone and of gradient type, which both are subclasses of the class of pre-monotone type. We briefly discuss the literature about initial-boundary value problems to constitutive equations, for which the resulting differential operator is monotone. To illustrate the classes introduced, in Section 3.3 we examine the models from Section 2.2 and show that they all belong to the class of pre-monotone type. For the first two of them we also show that they are of monotone-gradient type, but for the remaining seven engineering models the complicated investigation of whether they belong to one of the classes of gradient, monotone, or monotone-gradient type is left to later chapters.

The basis of our approach is the existence theory for constitutive equations of monotone type, and **Chapter 4** is thus devoted to the proof, that in the dynamic case under an additional coercivity assumption initial-boundary value problems to constitutive relations of monotone type have unique solutions. The proof is based on the general theory of evolution equations to maximal monotone operators.

The general investigation of the transformation of internal variables forms the content of **Chapter 5**. The vector $z = z(x,t) \in \mathbb{R}^N$ of internal variables is transformed to new internal variables $h = h(x,t) \in \mathbb{R}^N$ by

$$h(x,t) = H(z(x,t)),$$

with a vector field $H : \mathbb{R}^N \to \mathbb{R}^N$, which we call the transformation field. This transformation field induces a transformation of constitutive equations: the original constitutive equation, that is, the system of differential equations satisfied by the original internal variables z, is transformed into the new constitutive equation, the system of differential equations satisfied by h. We define new classes of constitutive equations, that consist of those constitutive equations to which transformation fields can be found, such that the transformed equations belong to one of the classes of pre-monotone, gradient, monotone or monotone-gradient type, prove some general results for this transformation, and finally state criteria that characterize the newly defined classes. There is a certain hierarchy of these criteria, since the gradient and monotone classes are subclasses of the pre-monotone class, which means that constitutive equations transformable to one of the first two classes must in particular satisfy all the criteria guaranteeing that transformation to the pre-monotone class is possible and at least one additional criterion.

Most of these criteria are not very explicit though, since to verify them it is necessary to know the transformation field. Here the problem arises of finding the transformation field which performs the desired transformation of constitutive equations into equations of a given class. We only have partial solutions to this problem. What we can show in Section 5.4 is that the additional criterion which must be satisfied by constitutive equations transformable to gradient type is that a transformation field H exists satisfying a nonlinear system of first order partial differential equations of the form

$$H'(z)^T H'(z) f(z) = \nabla \varphi(z),$$

where $f : \mathbb{R}^N \to \mathbb{R}^N$ is a function appearing in the given constitutive equations, $H'(z)$ is the Jacobian matrix of H and $H'(z)^T$ denotes the transpose of $H'(z)$. In this case the transformation field can be determined as the solution of this system of partial differential equations.

The appearance of this system of partial differential equations indicates that H is connected with the introduction of a new metric. To obtain solutions of this system the Cauchy-Kowalevski theorem can be applied, but in this work we do not study the existence theory for this system. In the additional criterion for constitutive equations transformable to monotone type this system of N partial differential equations is replaced by a system of N inequalities, which must be satisfied by the components of H and their derivatives up to second order, and which expresses that a certain matrix must be negative semi-definite. This criterion is stated in condition (v) of Theorem 5.3.1.

It turns out that those criteria which guarantee that the constitutive equations can be transformed to pre-monotone type and which in addition to the solvability of the system of partial differential equations must be satisfied by constitutive equations transformable to gradient type, restrict the form of the transformation field H. This means that only solutions of the system of partial differential equations with a special form are allowed as transformation fields. Solutions with this special form exist only if the coefficients of the system of partial differential equations, which are essentially given by the function f from the constitutive equation, satisfy compatibility conditions. Therefore these compatibility conditions for f determine the class of constitutive equations which can be transformed to gradient type, and thus characterize this class. We call them *classification conditions*.

Chapter 6 is devoted to the study of these classification conditions. More precisely, in this chapter the transformation of constitutive equations of the pre-monotone class to equations of the gradient class is investigated, and those classification conditions are derived which characterize the subclass of the pre-monotone class consisting of equations transformable to gradient type. This problem is of interest, firstly, since almost all constitutive equations from mechanics and the engineering sciences are of pre-monotone type, and secondly, since transformation to gradient type often automatically yields equations of monotone-gradient type, for which an existence theory is available.

In **Chapter 7** the transformation of rate independent constitutive relations is discussed. The theory runs in parallel to the theory for rate dependent relations but is slightly simpler, since rate independent relations are strongly structured. In particular, rate independent constitutive relations of monotone type are automatically of monotone-gradient type. An example in Section 8.5 shows that this property is not shared by rate dependent relations. The chapter concludes with two examples.

As an example of the transformation of rate dependent constitutive equations and of the application of the theory developed in Chapters 5 and 6 to engineering models, in **Chapter 8** one of the engineering models introduced in Section 2.2 is studied in detail.

The investigations of Chapter 8 show that the transformation of interior variables enables us to consider problems from engineering for which the theory of evolution equations for monotone operators originally could not be applied. On the other hand, they also show that many constitutive equations yield initial-boundary value problems, which do not even belong to this enlarged class, and therefore that the theory, as it is developed in this work, is not general enough to furnish a complete existence theory for all initial-boundary value problems of inelastic material behavior of solids in the case of small deformations. One can express this by saying that transformation of interior variables enlarges the class of problems, to which the theory of monotone operators is applicable, to an extent, which brings engineering problems into the 'boundary region' of the enlarged class.

Stated briefly, the class of constitutive equations transformable to monotone type is still too small. In **Chapter 9** we point out that the main restriction of this class comes from the form of the transformation fields we start out with, and how by the choice of a larger class of transformation fields a more general transformation theory might be developed which could be applied to many more constitutive equations. Though we believe that this generalization is possible, severe problems remain which must be overcome.

In Chapter 9 we also discuss some other topics from the recent literature about the initial-boundary value problems studied in this work and about hysteresis operators.

Finally, in the **Appendix** we show how the dissipation inequality, which is stated in Chapter 2 and which mainly motivates the definition of the class of generalized standard materials, is derived from the second law of thermodynamics. This appendix is also helpful to understand the formulation of the initial-boundary value problems in Chapter 2.

Chapter 2

Initial-Boundary Value Problems for the Inelastic Behavior of Metals

In this chapter we formulate the initial-boundary value problems which are studied in this work and subsequently present several examples from the literature in solid mechanics and engineering sciences of the constitutive equations appearing in these problems. The considerations about thermodynamics given in the appendix are helpful to understand the formulation of these initial-boundary value problems, but for a complete discussion of the mechanical background, which leads to their formulation, we must refer the reader to the mechanics literature. Some references to the literature are given at the end of the appendix.

2.1 Formulation of the Initial-Boundary Value Problems

Let S^3 denote the set of symmetric 3×3 matrices. Sometimes we consider elements of S^3 also to be elements of $\mathbb{R}^9$ and thus identify S^3 with a six dimensional subspace of $\mathbb{R}^9$. For 3×3 matices σ and τ we define the scalar product

$$\sigma \cdot \tau = \sum_{i,j=1}^{3} \sigma_{ij} \tau_{ij} \, ,$$

which coincides with the scalar product of σ and τ when considered to be elements of $\mathbb{R}^9$. In this work we adopt the convention to consider the gradient $\nabla \varphi$ of a real valued function $\varphi : \mathbb{R}^\ell \to \mathbb{R}$ as column vector. The transpose of a matrix or linear mapping A is denoted by A^T. We do not distinguish in notation between the transpose and the dual of a linear mapping and thus A^T also denotes the dual of A.

Let $\Omega \subseteq \mathbb{R}^3$ be an open set with smooth boundary $\partial\Omega$. This set represents the material points of a solid body. Let $u(x,t) \in \mathbb{R}^3$ be the displacement of the material point $x \in \Omega$ at time $t \geq 0$, let $\rho > 0$ be the density, which we assume to be constant in space and time, and let $T(x,t) \in S^3$ be the Cauchy stress tensor. We study the problem

$$\rho u_{tt}(x,t) = \mathrm{div}_x \, T(x,t) \tag{2.1.1}$$

$$T(x,t) = D(\varepsilon(\nabla_x u(x,t)) - \varepsilon_p(x,t)) \qquad (2.1.2a)$$

$$z_t(x,t) = f(\varepsilon(\nabla_x u(x,t)), z(x,t))\,. \qquad (2.1.3a)$$

Here the Jacobi matrix $\nabla_x u$ and the divergence $\operatorname{div}_x T$ are defined by

$$\nabla_x u(x,t) = \Big(\frac{\partial}{\partial x_j}\, u_i(x,t)\Big)_{i,j=1,2,3}$$

$$\operatorname{div}_x T(x,t) = \Big(\sum_{j=1}^{3} \frac{\partial}{\partial x_j}\, T_{ij}(x,t)\Big)_{i=1,2,3} \in \mathbb{R}^3$$

and the subscript t denotes differentiation with respect to time. Moreover

$$\varepsilon(\nabla_x u(x,t)) = \frac{1}{2}(\nabla_x u(x,t) + (\nabla_x u(x,t))^T) \in \mathcal{S}^3 \qquad (2.1.4)$$

is the linear strain tensor,

$$\varepsilon_p(x,t) \in \mathcal{S}^3$$

is the plastic strain tensor, which is also a symmetric matrix, and

$$z(x,t) = (\varepsilon_p(x,t), \tilde{z}(x,t)) \in \mathcal{S}^3 \times \mathbb{R}^{N-9} \subseteq \mathbb{R}^N$$

is the vector of internal variables. The first nine components of z are given by the elements of ε_p, which also belong to the internal variables, hence $\tilde{z}(x,t) \in \mathbb{R}^{N-9}$.

$D : \mathcal{S}^3 \to \mathcal{S}^3$ denotes a symmetric linear mapping, which is called elasticity tensor. In this work we assume that the mapping D is positive definite, and that D is independent of x and t. Finally

$$f : D(f) \subseteq \mathcal{S}^3 \times \mathbb{R}^N \to \mathbb{R}^N$$

is a given function, where $D(f)$ denotes the domain of f. This function together with the special choice of the elasticity tensor D defines the material behavior. (2.1.2a) and (2.1.3a) are constitutive equations, whereas (2.1.1) is the conservation law of linear momentum.

We assume that one of the two homogeneous boundary conditions

$$u(x,t)\big|_{x \in \partial\Omega} = 0 \qquad (2.1.5)$$

or

$$T(x,t)n(x)\big|_{x \in \partial\Omega} = 0 \qquad (2.1.6)$$

is satisfied, where $n(x)$ denotes the exterior unit normal vector to $\partial\Omega$ at $x \in \partial\Omega$. The initial conditions are

$$u(x,0) = u^{(0)}(x)\,,\ u_t(x,0) = u^{(1)}(x)\,,\ z(x,0) = z^{(0)}(x)\,,\quad x \in \Omega\,, \qquad (2.1.7)$$

with given functions

$$u^{(0)}, u^{(1)} : \Omega \to \mathbb{R}^3\,,\quad z^{(0)} : \Omega \to \mathbb{R}^N\,.$$

We also require that there exists a function $\psi : D(f) \subseteq \mathcal{S}^3 \times \mathbb{R}^N \to [0, \infty)$ with

$$\rho \nabla_\varepsilon \psi(\varepsilon, z) = T \qquad (2.1.8)$$

$$\rho \nabla_z \psi(\varepsilon, z) \cdot f(\varepsilon, z) \le 0 \qquad (2.1.9)$$

for all $(\varepsilon, z) \in D(f)$. The function ψ is called free energy, and (2.1.9) is called dissipation inequality. The requirement, that such a function ψ exists, is a consequence of the second law of thermodynamics. This is shown in the appendix, where the derivation of (2.1.8) and (2.1.9) is given. Note that the existence of a free energy satisfying (2.1.8) and (2.1.9) is solely a requirement for f and restricts the form of the constitutive equation (2.1.3a), since from (2.1.2a) and from the symmetry of D it follows that (2.1.8) holds, if and only if

$$\rho \psi(\varepsilon, z) = \frac{1}{2} \left[D(\varepsilon - \varepsilon_p) \right] \cdot (\varepsilon - \varepsilon_p) + \psi_1(z)$$

with a suitable function $\psi_1 : D(\psi_1) \subseteq \mathbb{R}^N \to [0, \infty)$. This function is obtained as constant of integration when (2.1.8) is integrated with respect to ε. Therefore to the given function f there must exist ψ_1 such that the dissipation inequality (2.1.9) holds. The constitutive equations (2.1.2a) and (2.1.3a) are called thermodynamically admissible if a free energy exists satisfying (2.1.8) and (2.1.9).

The relations (2.1.1) – (2.1.9) comprise the initial-boundary value problem to be studied. This formulation of the problem is restrictive, however, since it does not include the important class of rate independent constitutive models. We now define this class and then give a more general formulation, which includes models of this class.

The constitutive equation (2.1.3a) assigns to given initial data $z^{(0)}$ and a given function $\varepsilon(x, t) = \varepsilon(\nabla_x u(x, t))$ the function $z : \Omega \times [0, \infty) \to \mathbb{R}^N$, and thus defines the constitutive relation

$$z(x, t) = \mathop{\mathcal{F}}_{0 \le \tau \le t} (\varepsilon(x, \tau), z^{(0)}(x)). \qquad (2.1.10)$$

Definition 2.1.1 *We call the constitutive relation* (2.1.10) *rate independent if for all* $z, \varepsilon, z^{(0)}$ *satisfying* (2.1.10) *we have*

$$z(x, \lambda t) = \mathop{\mathcal{F}}_{0 \le \tau \le t} (\varepsilon(x, \lambda t), z^{(0)}(x)),$$

for all $\lambda \ge 0$.

A constitutive relation is therefore rate independent if it is invariant under scaling of the time variable t. The differential equation (2.1.3a) is scaling invariant only in the degenerate case $f(\varepsilon, z) \equiv 0$, and therefore always defines a rate dependent model, but it can be considered to be a special case of the more general constitutive relation

$$z_t(x, t) \in f\big(\varepsilon(\nabla_x u(x, t)), z(x, t)\big), \qquad (2.1.11)$$

which also includes rate independent constitutive models. Here

$$f : D(f) \subseteq \mathcal{S}^3 \times \mathbb{R}^N \to \mathcal{P}(\mathbb{R}^N)$$

is a mapping with values in the set $\mathcal{P}(\mathbb{R}^N)$ of all subsets of $\mathbb{R}^N$. The domain of f is as usual defined by

$$D(f) = \{(\varepsilon, z) \in \mathcal{S}^3 \times \mathbb{R}^N \mid f(\varepsilon, z) \neq \emptyset\}.$$

The relation (2.1.11) is a generalization of (2.1.3a), since the equation (2.1.3a) can be written in the form (2.1.11) by replacing f from (2.1.3a) with the function with values in $\mathcal{P}(\mathbb{R}^N)$ that assigns to (ε, z) the value $\{f(\varepsilon, z)\}$. Such functions are called single valued. The following lemma shows that (2.1.11) includes rate independent constitutive models:

Lemma 2.1.2 *Suppose that for every* $(\varepsilon, z) \in D(f)$, *whenever* $\zeta \in \mathbb{R}^N$ *is in* $f(\varepsilon, z)$, *the set* $f(\varepsilon, z)$ *also contains all elements* $\lambda\zeta$ *with* $\lambda \geq 0$. *Then* (2.1.11) *is rate independent. Conversely, let* (ε, z) *be a pair of functions such that* $(x, t) \mapsto (\varepsilon(x, \lambda t), z(x, \lambda t))$ *solves* (2.1.11) *for all* $\lambda \geq 0$. *Then*

$$\{\lambda\zeta \mid \lambda \geq 0\} \subseteq f(\overline{\varepsilon}, \overline{z})$$

for all elements $(\zeta, \overline{\varepsilon}, \overline{z})$ *from the range of the triple of functions* (z_t, ε, z).

We leave the obvious proof of this lemma to the reader. The first statement shows that if every value of f is a conic subset of $\mathbb{R}^N$ with vertex at 0, then the constitutive relation is rate independent, whereas the second statement is almost the converse: If (2.1.11) is rate independent, then every value of f must contain a cone, if this value is really relevant, that is, if it contains an element from the range of the time derivative z_t of a solution z.

A generalization of the initial-boundary value problem formulated above is thus obtained by replacing the constitutive equation (2.1.3a) by the constitutive relation (2.1.11). We also generalize equation (2.1.2a) and replace it by

$$T(x, t) = D(\varepsilon(\nabla_x u(x, t)) - Bz(x, t)), \tag{2.1.12}$$

where $B : \mathbb{R}^N \to \mathcal{S}^3$ is a given linear operator. The reason for this generalization will become obvious in the following section when we discuss examples and in Chapter 5 when we discuss the transformation of constitutive equations. Since the symmetric matrix ε_p in (2.1.2a) consists of the first nine components of z, we obtain (2.1.2a) as a special case of (2.1.12) if, for example, we define B by

$$z \mapsto Bz = \varepsilon(z_p) = \frac{1}{2}(z_p + z_p^T),$$

where z_p denotes the 3×3 matrix formed out of the first nine components of z. The general initial-boundary value problem which we study in this work thus consists of the equations

$$\rho u_{tt} = \operatorname{div}_x T \tag{2.1.1}$$

$$T = D(\varepsilon(\nabla_x u) - Bz) \tag{2.1.2}$$

$$z_t \in f(\varepsilon(\nabla_x u), z) \tag{2.1.3}$$

together with the equations and relations (2.1.4) – (2.1.9), where (2.1.9) is now interpreted to mean

$$\rho \nabla_z \psi(\varepsilon, z) \cdot \zeta \leq 0$$

for every $\zeta \in f(\varepsilon, z)$. From (2.1.8), (2.1.2) and from the symmetry of D we now obtain

$$\rho\psi(\varepsilon, z) = \frac{1}{2}[D(\varepsilon - Bz)] \cdot (\varepsilon - Bz) + \psi_1(z) \qquad (2.1.13)$$

with a suitable function $\psi_1 : D(\psi_1) \subseteq \mathbb{R}^N \to [0, \infty)$.

2.2 Examples of Constitutive Equations

We next present examples of constitutive equations. The first two examples are the classical and basic models of Prandtl-Reuss and Maxwell. They are followed by seven examples of engineering models which were developed in the last twenty years. Here we state the model equations and give some additional motivating remarks, but do not analyse the equations. In particular we do not discuss whether the model equations are thermodynamically admissible. In several sections of this work these constitutive models are used to illustrate the theory developed there, and in this context we also show for several of these models that they are thermodynamically admissible.

The models we selected are under discussion in the engineering and mechanics communities, but their selection is not motivated by their usefulness in the theory of metals, which we are not able to judge, but rests on the limited and partly accidental knowledge of the author. In fact, we are interested in developing a mathematical theory for a class of constitutive equations as large as possible, which ideally should also include less useful constitutive models.

Example 1. The first example is the constitutive model of Prandtl-Reuss, which is rate independent. This is the constitutive law studied most often in the literature, and accounts of it can be found in [62, 107, 146, 156, 183, 191, 201, 206], for example. For this model the equations (2.1.1) – (2.1.3) have the form

$$\rho u_{tt} = \mathrm{div}_x\, T$$
$$T = D(\varepsilon(\nabla_x u) - \varepsilon_p) \qquad (2.2.1)$$
$$\frac{\partial}{\partial t}\varepsilon_p \in g_K(T), \qquad (2.2.2)$$

where $\varepsilon_p \in \mathcal{S}^3$ is the plastic strain tensor, and where with a non-empty, closed, convex subset K of $\mathcal{S}^3$ the function $g_K : \mathcal{S}^3 \to \mathcal{P}(\mathcal{S}^3)$, which defines the flow rule, is given by

$$g_K(\sigma) = \begin{cases} \{n \in \mathcal{S}^3 \mid \forall \tau \in \mathcal{S}^3 : n \cdot (\sigma - \tau) \le 0\}, & \sigma \in K \\ \emptyset, & \sigma \in \mathcal{S}^3 \backslash K. \end{cases} \qquad (2.2.3)$$

This definition implies that $g_K(\sigma)$ is a conic subset of $\mathcal{S}^3$ for all $\sigma \in \mathcal{S}^3$, that $g_K(\sigma) = \{0\}$ for $\sigma \in \overset{\circ}{K}$, and that for $\sigma \in \partial K$ the set $g_K(\sigma)$ consists of all multiples $\lambda n(\sigma)$ of the exterior unit normal vector $n(\sigma)$ to ∂K at σ with $\lambda \ge 0$, if the boundary ∂K is smooth in a neighborhood of σ. The convex set K is specified by the yield criterion. For the von Mises yield criterion [150, 62] this set is

$$K = \{\sigma \in \mathcal{S}^3 \mid (P_0\sigma) \cdot (P_0\sigma) \le c\}$$

where $c > 0$ is a given constant, and where $P_0 : S^3 \to S^3$ is the orthogonal projection onto the subspace $\{\sigma \in S^3 \mid \text{trace}(\sigma) = 0\}$. A short computation shows that P_0 is given by

$$P_0 \sigma = \sigma - \frac{1}{3} \text{trace}(\sigma) I \,,$$

where I denotes the identity on S^3. If T denotes the stress, then $P_0(T)$ is called stress deviator.

For Tresca's yield criterion [210, 62] the convex set is defined by

$$K = \{\sigma \in S^3 \mid \sup_{1 \le i,j \le 3} |\sigma_i - \sigma_j| \le c\}\,,$$

where σ_i are the eigenvalues of $\sigma \in S^3$. It is shown in [62, p.232] that this set K is convex.

To write (2.2.1), (2.2.2) in the form of the equations (2.1.2), (2.1.3), we can set $N = 9$, choose for the vector of internal variables $z(x,t) = \varepsilon_p(x,t) \in \mathbb{R}^N$, select for the linear mapping $B : \mathbb{R}^N \to S^3$ any projector of $\mathbb{R}^N$ onto the subspace S^3 of $\mathbb{R}^N$, and define $f : S^3 \times \mathbb{R}^N \to \mathcal{P}(S^3) \subseteq \mathcal{P}(\mathbb{R}^N)$ by

$$f(\varepsilon, z) = g_K(T) = g_K(D(\varepsilon - Bz))\,.$$

As initial data for z we must choose a function $z^{(0)} = \varepsilon_p^{(0)}$, whose values are all contained in S^3.

Since $Bz = z$ for $z \in S^3$, it is immediately seen that with f and B defined in this way the relations (2.1.2), (2.1.3) are equivalent to (2.2.1), (2.2.2). However, these definitions have introduced too many internal variables. For these f and B the 'history functional' $(\varepsilon(\nabla_x u(x, \cdot)), z^{(0)}(x)) \mapsto T(x, \cdot)$ defined by (2.1.2), (2.1.3) is not at all influenced by the component of $z \in \mathbb{R}^9$ contained in the complementary subspace $\ker(B)$ of S^3. The existence of this unnecessary non-trivial subspace $\ker(B)$ often needs special consideration and in many investigations introduces technical difficulties. It is therefore more convenient to avoid the introduction of unnecessary internal variables by choosing for the linear mapping B an isomorphism between $\mathbb{R}^6$ and the six-dimensional subspace S^3 of $\mathbb{R}^9$, in which case we can set $N = 6$. The vector $z = \varepsilon_p$ of internal variables is then an element of $\mathbb{R}^6$, and $f(\varepsilon, z) = B^{-1} g_K(D(\varepsilon - Bz))$ is defined on the space $S^3 \times \mathbb{R}^6$ and has range $\mathcal{P}(\mathbb{R}^6)$. Here $B^{-1} : \mathcal{P}(S^3) \to \mathcal{P}(\mathbb{R}^6)$ is given by

$$B^{-1}V = \{B^{-1}\sigma \mid \sigma \in V\}\,.$$

This is the point of view in which we choose to study this model and other models throughout this work. To get a concise notation in these studies we do not explicitly denote the isomorphism between S^3 and $\mathbb{R}^6$. In other words, we simply identify S^3 with $\mathbb{R}^6$ and, for example, write $f(\varepsilon, z) = g_K(D(\varepsilon - z))$.

That the model of Prandtl-Reuss is rate independent follows from Lemma 2.1.2, since $f(\varepsilon, z)$ satisfies the condition of this lemma for every (ε, z). The rheological diagram for this model is displayed in figure 2.2.1.

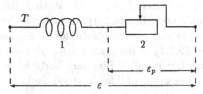

1 : Hookeian element

2 : Dry-friction element (rate independent)

Figure 2.2.1

In this figure the element 2 symbolizes a dry-friction element, whose rate independent behavior is described by the constitutive equation (2.2.2), while the symbol 1 is a Hookeian element, whose behavior is described by equation (2.2.1).

Example 2. If one replaces the dry-friction element in figure 2.2.1 by a dashpot, then one obtains a rate dependent constitutive model, the classical Maxwell model, with the rheological diagram shown in figure 2.2.2.

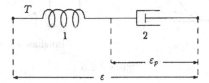

1 : Hookeian element

2 : Dashpot (rate dependent)

Figure 2.2.2

For this model the equations (2.1.1) – (2.1.3) become

$$\rho u_{tt} = \operatorname{div}_x T$$
$$T = D(\varepsilon(\nabla_x u) - \varepsilon_p)$$
$$\frac{\partial}{\partial t} \varepsilon_p = g(T), \qquad (2.2.4)$$

where $\varepsilon_p \in \mathcal{S}^3$ is again the plastic strain tensor, or better, since the model is rate dependent, the viscous strain tensor, and where the function $g : \mathcal{S}^3 \to \mathcal{S}^3$ is typically of the form

$$g(T) = \Gamma(|P_0 T|) \frac{P_0 T}{|P_0 T|}$$

with $|P_0 T| = (P_0 T \cdot P_0 T)^{1/2}$ and with a given function $\Gamma : [0, \infty) \to [0, \infty)$. One often chooses

$$\text{(i)} \quad \Gamma(\xi) = C\xi^n \quad \text{or} \quad \text{(ii)} \quad \Gamma(\xi) = C\langle \xi - k \rangle^n$$

with

$$\langle \xi \rangle = \max(\xi, 0).$$

$C > 0$, $k > 0$ and $n > 0$ are material constants. With Γ from (i) the constitutive equation (2.2.4) is Norton's law. In rheology, a material described by this law is called dilatant for $n \in (0, 1]$ and pseudo-plastic for $n \in [1, \infty)$. With Γ from (ii) the constitutive equation (2.2.4) is Bingham's law. A body, whose behavior can be modeled by Bingham's law is called Bingham body for $n = 1$ and generalized Bingham body for $n \neq 1$. The graph of Γ for Norton's and Bingham's law is depicted in figure 2.2.3.

To model the behavior of metals both Norton's and Bingham's law are used with the exponent usually chosen to be $n \sim 3 \ldots 9$. If $|T|$ is sufficiently small, then for such n the value of $\Gamma(|P_0 T|)$ is very small. Hence $|\frac{\partial}{\partial t} \varepsilon_p| = |g(T)|$ is very small. This is the region where the material displays almost elastic behavior. In the region of stress, where the graph of $\Gamma(|P_0 T|)$ is strongly increasing, $|\frac{\partial}{\partial t} \varepsilon_p| = |g(T)|$ is large, which means that the material exhibits viscous flow.

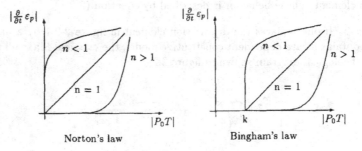

Figure 2.2.3

For this example the equation (2.2.4) can be written in the form of (2.1.3) if we set

$$f(\varepsilon, \varepsilon_p) = g(T) = g(D(\varepsilon - \varepsilon_p)) = \Gamma(|P_0 T|) \frac{P_0 T}{|P_0 T|}$$

with $f : D(f) = \mathcal{S}^3 \times \mathcal{S}^3 \cong \mathcal{S}^3 \times \mathbb{R}^N \to \mathcal{S}^3 \cong \mathbb{R}^N$, where we again choose $N = 6$ and identify $\mathcal{S}^3$ with $\mathbb{R}^6$.

More complicated constitutive models are obtained by combinations of Hookeian elements and dashpots. A typical model is displayed in figure 2.2.4.

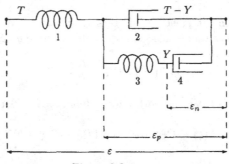

Figure 2.2.4

In this example $\varepsilon_p, \varepsilon_n \in \mathcal{S}^3$ have the dimension of strain and $Y \in \mathcal{S}^3$ the dimension of stress. The constitutive equation of the linear spring 3 is

$$Y = \mathcal{M}(\varepsilon_p - \varepsilon_n) \qquad (2.2.5)$$

with a positive scalar $\mathcal{M}$, and the constitutive equations for the dashpots 2 and 4 are

$$\frac{\partial}{\partial t}\varepsilon_p = \hat{g}(T - Y) = \hat{\Gamma}(|P_0(T - Y)|)\frac{P_0(T - Y)}{|P_0(T - Y)|} \qquad (2.2.6)$$

$$\frac{\partial}{\partial t}\varepsilon_n = \tilde{g}(Y) = \tilde{\Gamma}(|Y|)\frac{Y}{|Y|}, \qquad (2.2.7)$$

with given functions $\hat{\Gamma}, \tilde{\Gamma} : [0, \infty) \to [0, \infty)$ satisfying $\hat{\Gamma}(0) = \tilde{\Gamma}(0) = 0$. Typically the graphs of $\hat{\Gamma}$ and $\tilde{\Gamma}$ are similar to the one displayed in figure 2.2.3.

Combining (2.2.5) – (2.2.7) we obtain

$$\rho u_{tt} = \operatorname{div}_x T \qquad (2.2.8)$$

$$T = D(\varepsilon(\nabla_x u) - \varepsilon_p) \qquad (2.2.9)$$

$$\frac{\partial}{\partial t}\varepsilon_p = \hat{g}\big(D(\varepsilon(\nabla_x u) - \varepsilon_p) - \mathcal{M}(\varepsilon_p - \varepsilon_n)\big) \qquad (2.2.10)$$

$$\frac{\partial}{\partial t}\varepsilon_n = \tilde{g}(\mathcal{M}(\varepsilon_p - \varepsilon_n)). \qquad (2.2.11)$$

To write these equations in the form of the equations (2.1.1) – (2.1.3) we identify $\mathcal{S}^3$ with $\mathbb{R}^6$, set $N = 12$, choose for the vector of internal variables $z = (\varepsilon_p, \varepsilon_n) \in \mathcal{S}^3 \times \mathcal{S}^3 \cong \mathbb{R}^N$, and define

$$f : D(f) = \left(\mathcal{S}^3\right)^3 \cong \mathcal{S}^3 \times \mathbb{R}^N \to \left(\mathcal{S}^3\right)^2 \cong \mathbb{R}^N$$

by

$$f(\varepsilon, z) = f(\varepsilon, \varepsilon_p, \varepsilon_n) = g\big(D(\varepsilon - \varepsilon_p) - \mathcal{M}(\varepsilon_p - \varepsilon_n), -\mathcal{M}(\varepsilon_p - \varepsilon_n)\big)$$
$$= \big(\hat{g}(D(\varepsilon - \varepsilon_p) - \mathcal{M}(\varepsilon_p - \varepsilon_n)), \tilde{g}(\mathcal{M}(\varepsilon_p - \varepsilon_n))\big).$$

The reason for the minus sign in the second argument of g will become clear in Section 3.3 where the classification of this model is discussed. Of course, $B : \mathcal{S}^3 \times \mathcal{S}^3 \to \mathcal{S}^3$ is defined by $Bz = B(\varepsilon_p, \varepsilon_n) = \varepsilon_p$. With these definitions of f and B the equations (2.1.1) – (2.1.3) are equivalent to (2.2.8) – (2.2.11).

Y is called backstress. It is a variable of kinematic hardening, and defines in the stress space the center of elastic behavior of the material. A mathematical investigation of such models can be found in [16, 17, 133], for example.

Example 3. We next consider the constitutive model proposed by J.L. Chaboche. A number of modifications of the equations of this model have been published, cf. for example [29, 30, 31, 33, 34, 63]. The equations we state are taken from [63].

For this model the rheological diagram is also given by figure 2.2.4, but dashpot 2 now displays isotropic hardening. The equations (2.1.1) – (2.1.3) take the form

$$\rho u_{tt} = \text{div}_x\, T \tag{2.2.12}$$

$$T = D(\varepsilon(\nabla_x u) - \varepsilon_p) \tag{2.2.13}$$

$$\frac{\partial}{\partial t}\,\varepsilon_p = \left\langle \frac{|P_0(T-Y)| - \varsigma}{C_1} \right\rangle^n \frac{P_0(T-Y)}{|P_0(T-Y)|} \tag{2.2.14}$$

$$\frac{\partial}{\partial t}\,\varepsilon_n = C_2 |Y|^m \frac{Y}{|Y|} + C_3 \left| \frac{\partial}{\partial t}\,\varepsilon_p \right| Y \tag{2.2.15}$$

$$\frac{\partial}{\partial t}\,\varsigma = C_4 \left| \frac{\partial}{\partial t}\,\varepsilon_p \right| (\varsigma_0 - \varsigma) \tag{2.2.16}$$

with the additional equation

$$Y = \mathcal{M}(\varepsilon_p - \varepsilon_n). \tag{2.2.17}$$

Here, as above, $\langle \xi \rangle = \max(\xi, 0)$, and $m, \mathcal{M}, n, C_1, \varsigma_0 > 0$, $C_2, C_3, C_4 \geq 0$ are material parameters. $\varsigma(x,t)$ is the variable of isotropic hardening. It determines the size of the elastic region in the stress space. This region is bounded by the surface $|P_0(T-Y)| = \varsigma$. Equation (2.2.16) is the evolution equation for ς. The initial data are chosen such that

$$0 < \varsigma(x,0) \leq \varsigma_0$$

for all x. Equation (2.2.16) then implies $\varsigma(x,0) \leq \varsigma(x,t) \leq \varsigma_0$.

Equation (2.2.15) is the constitutive equation for dashpot 4 in figure 2.2.4. This equation differs in an essential way from the constitutive equation (2.2.7) for dashpot 4 in the preceding example, since the last term on the right hand side of equation (2.2.15) depends on the flow rate $|\frac{\partial}{\partial t}\,\varepsilon_p| = \left\langle \frac{|P_0(T-Y)| - \varsigma}{C_1} \right\rangle^n$. Already here we remark that from the point of view of the mathematical theory developed in the following sections this term introduces difficulties. Without this term the constitutive model of Chaboche belongs to the class $\mathcal{TM}$ of constitutive equations, which can be transformed to monotone type. This class is introduced in Section 5.

In [63] values for the parameters in this model were given, which were chosen to adapt the model to the mechanical properties of INCONEL 718, a nickel-based super-alloy primarily comprised of approximately 52 percent nickel, 18 percent chromium, 18 percent iron, and 5 percent columbium and tantalum. The exponents in (2.2.14) and (2.2.15) were chosen as $n = 5.10$ and $m = 7.00$.

A modification of this model is given by Nouailhas [169] (cf. also [21]). In this modification the number of internal variables is larger.

Example 4. The following model of Bodner and Partom [18, 200, 63] is used in Chapter 8 to illustrate the application of the transformation theory developed in this work to constitutive equations, and thus is discussed there in detail. We present here a simple version of the model equations, where as in the original form of this model, kinematic hardening is not taken into account, but only isotropic hardening. The rheological diagram is therefore given by figure 2.2.2, with dashpot 2 displaying isotropic hardening. In this model the equations (2.1.1) – (2.1.3) take the form

$$\rho u_{tt} = \text{div}_x\, T \tag{2.2.18}$$

$$T = D(\varepsilon(\nabla_x u) - \varepsilon_p) \tag{2.2.19}$$

$$\frac{\partial}{\partial t}\, \varepsilon_p = d\, \exp\left(-\alpha\left(\frac{\zeta}{|P_0 T|}\right)^n\right) \frac{P_0 T}{|P_0 T|} \tag{2.2.20}$$

$$\frac{\partial}{\partial t}\, \zeta = m(\zeta_1 - \zeta)\left|\frac{\partial}{\partial t}\, \varepsilon_p\right| |P_0 T| - A\zeta_1\left(\frac{\zeta - \zeta_2}{\zeta_1}\right)^r \tag{2.2.21}$$

with material parameters $n, r > 1$, $\alpha, d, m > 0$, $A \geq 0$, $\zeta_1 > \zeta_2 > 0$. The variable of isotropic hardening is $\zeta(x,t)$. The initial data are chosen such that $\zeta_2 \leq \zeta(x,0) \leq \zeta_1$ for all x. The evolution equation (2.2.21) then implies that $\zeta_2 \leq \zeta(x,t) \leq \zeta_1$ for all (x,t). Values for the parameters of this model to adapt it to the mechanical properties of INCONEL 718 can also be found in [63]. The values given there for the exponents are $n = 6.0$, $r = 7.0$. If A is not zero then the last term on the right hand side is negative. This term models static recovery of the material.

The model of Bodner and Partom differs from most others since the right hand side of (2.2.20) is a bounded function of T and ζ. However, for small and medium values of $|T|$ the equation (2.2.20), which is the constitutive equation of the dashpot in figure 2.2.2, is of the usual form and differs little from other models. From the data given in [63] we surmise that a difference occurs only for values of $|T|$ which are larger than values in normal engineering practice.

Existence, uniqueness, regularity and asymptotic behavior of solutions to initial-boundary value problems for this constitutive model were investigated in [36, 37, 38, 39, 40, 43].

Example 5. In [90] E.W. Hart proposed a constitutive model with rheological diagram given in figure 2.2.4. In this model the dashpot 4, simply called 'element' by Hart, has an unusual constitutive equation. Moreover, this model differs from that of Chaboche in that element 4 displays isotropic hardening instead of dashpot 2.

For the model of Hart the equations (2.1.1) – (2.1.3) have the form

$$\rho u_{tt} = \text{div}_x\, T$$

$$T = D(\varepsilon(\nabla_x u) - \varepsilon_p)$$

$$\frac{\partial}{\partial t}\, \varepsilon_p = C_1 |P_0(T - Y)|^n\, \frac{P_0(T - Y)}{|P_0(T - Y)|}$$

$$\frac{\partial}{\partial t}\, \varepsilon_n = C_2\left(\frac{\zeta}{C_3}\right)^m \left(\ln\frac{\zeta}{|Y|}\right)^{-1/\lambda} \frac{Y}{|Y|} \tag{2.2.22}$$

$$\frac{\partial}{\partial t}\, \zeta = \gamma(\zeta, |Y|)\left|\frac{\partial}{\partial t}\, \varepsilon_n\right| |Y| \tag{2.2.23}$$

with the additional equation

$$Y = \mathcal{M}(\varepsilon_p - \varepsilon_n).$$

$C_1, \ldots, C_3$ are temperature dependent positive constants, and $\zeta(x,t) > 0$ is the variable of isotropic hardening. $1/\lambda$ is a positive constant, and the functions ζ and

Y must satisfy $|Y(x,t)| < \zeta(x,t)$ for almost all (x,t). The initial data are chosen such that this inequality holds for $t = 0$. For $t > 0$ it must be guaranteed by the evolution equations (2.2.22), (2.2.23), but it is not at all obvious that solutions of these evolution equations necessarily satisfy this inequality. This needs to be proved. The function γ is determined by fitting experimental data and thus depends on the metal considered. Hart notes that for 1100 Al at temperatures from one-third to more than one-half the melting temperature, a fair fit of experimental data could be made by the choice

$$\gamma(\zeta, |Y|) = C_4 \left(\frac{|Y|}{\zeta}\right)^k \left(\frac{\zeta}{C_3}\right)^{-m}$$

with a positive material constant C_4 and with $k = 7.8$. For the other exponents one has $n \sim 9$, $\lambda \sim 0.15$, $m = 4\ldots5$. Equation (2.2.22) is the constitutive equation for the element 4 in figure 2.2.4. The dependence of the right hand side of (2.2.22) on the absolute value of the backstress $|Y|$ with hardening parameter ζ fixed is shown in figure 2.2.5.

Hart notes that his equations should describe the nonelastic behavior of single crystal alloys and of polycrystalline materials in ranges of temperature and strain rate in which the grain boundaries contribute little to the observed strain rates. The regime thus covered is, for most technical alloys, the most familiar deformation range for temperatures up to about one-third the melting temperature.

A mathematical analysis of the initial-boundary value problem to this model is given in [1].

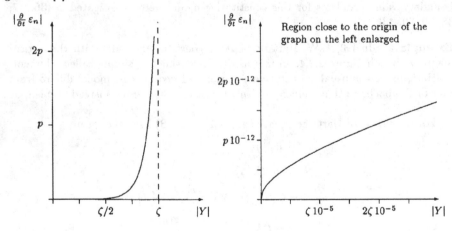

Figure 2.2.5

Example 6. In [147] A. Miller developed a model with rheological diagram given in figure 2.2.4. In this model dashpot 2 displays isotropic hardening. For this model the equations (2.1.1) – (2.1.3) take the form

$$\rho u_{tt} = \mathrm{div}_x\, T$$

$$T = D(\varepsilon(\nabla_x u) - \varepsilon_p)$$

$$\frac{\partial}{\partial t}\varepsilon_p = C_1\left[\sinh\left(\frac{|P_0(T-Y)|}{\zeta}\right)^{3/2}\right]^n \frac{P_0(T-Y)}{|P_0(T-Y)|}$$

$$\frac{\partial}{\partial t}\varepsilon_n = C_1\left[\sinh(A_1|Y|)\right]^n \frac{Y}{|Y|}$$

$$\frac{\partial}{\partial t}\zeta = C_2\left|\frac{\partial}{\partial t}\varepsilon_p\right|(C_3+|Y|-\frac{A_2}{A_1}\zeta^3) - C_1 C_2 C_3\left[\sinh(A_2\zeta^3)\right]^n$$

with the additional equation

$$Y = \mathcal{M}(\varepsilon_p - \varepsilon_n).$$

$\zeta(x,t) > 0$ is the variable of isotropic hardening, $\mathcal{M}, A_1, A_2, C_1, C_2, C_3, n$ are positive constants. In [148] the quantitative behavior of this model with type 304 stainless steel is demonstrated. The value for the exponent chosen in this investigation is $n = 5.8$. Existence and uniqueness of solutions to initial-boundary value problems for this model are studied in [110, 111].

Example 7. The model of O.T. Bruhns [26] (cf. also [21]) cannot be described by the rheological diagrams given above. It can be explained using the diagram in figure 2.2.6, but we warn the reader, since in this diagram the spring 3 is very unusual. It has a differential spring 'constant', which can be varied and controlled by an interior scalar variable $\zeta(x,t)$. This variable simultaneously controls the amount of kinematic and isotropic hardening. Kinematic hardening results from the dependence of this differential spring constant on ζ, whereas isotropic hardening results from the dependence of the yield surface on ζ.

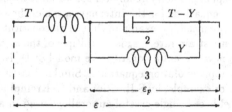

Figure 2.2.6

For this model the equations (2.1.1) – (2.1.3) take the form

$$\rho u_{tt} = \mathrm{div}_x T \tag{2.2.24}$$

$$T = D(\varepsilon(\nabla_x u) - \varepsilon_p) \tag{2.2.25}$$

$$\frac{\partial}{\partial t}\varepsilon_p = C_1\langle\lambda\rangle\left(1 + \frac{\lambda}{C_2}\right)^n \frac{P_0(T-Y)}{|P_0(T-Y)|} \tag{2.2.26}$$

$$\frac{\partial}{\partial t}Y = \frac{3}{2}\left(C_3 - \frac{1}{2}\kappa'(\zeta)\right)\frac{\partial}{\partial t}\varepsilon_p \tag{2.2.27}$$

$$\frac{\partial}{\partial t}\zeta = \sqrt{\frac{3}{2}}\kappa(\zeta)\left|\frac{\partial}{\partial t}\varepsilon_p\right| \tag{2.2.28}$$

with $\langle \lambda \rangle = \max(\lambda, 0)$ and with

$$\lambda = |P_0(T - Y)| - \sqrt{\frac{3}{2}}\,\kappa(\zeta),$$

$$\kappa(\zeta) = C_4 + 2C_3\left(\frac{C_6\zeta}{1 + C_5\zeta} + (1 - C_6)\,\frac{\ln(1 + \zeta)}{1 + C_5\ln(1 + \zeta)}\right).$$

$n, C_1, \ldots, C_5 > 0$ and $0 < C_6 < 1$ are material constants. $\zeta(x, t)$ is non-negative, and a short computation yields that $C_3 - \frac{1}{2}\kappa'(\zeta) > 0$ for all non-negative ζ. Equation (2.2.26) is the constitutive equation for dashpot 2, whereas (2.2.27) is the constitutive equation for the variable spring element 3. The differential spring constant is $\frac{3}{2}\left(C_3 - \frac{1}{2}\kappa'(\zeta)\right)$. The surface

$$\left\{T \in \mathcal{S}^3 \,\middle|\, |P_0(T - Y)| = \sqrt{\frac{3}{2}}\,\kappa(\zeta)\right\}$$

plays the role of a yield surface in the stress space. In [21] values of the material constants for the austenitic steel 1.4948 are given. In particular, the value for the exponent in (2.2.26) is $n = 5.33$.

Actually, the system of equations given above is a slightly simplified version of the model of Bruhns. In the general form the model contains yet another internal variable, which is introduced to better model cyclic effects. In the formulation given above we have omitted this variable.

Example 8. The equations in the preceding examples can describe anisotropic materials, since the elasticity tensor $D : \mathcal{S}^3 \to \mathcal{S}^3$ is only required to define a symmetric and positive definite mapping. Therefore any tensor for anisotropic materials can be inserted. The evolution equations for the internal variables, however, are assumed to be isotropic. By this we mean, for example, that in all constitutive models considered up to now the plastic strain rate $\frac{\partial}{\partial t}\varepsilon_p$ is a multiple of the stress tensor T or the overstress $T - Y$. We now present a constitutive model of D. Nouailhas and A.D. Freed [170] with anisotropic evolution equations. Similar constitutive models have been given previously, for example, by S.H. Choi and E. Krempl [45]. These models are used to study nickel-based single-crystal superalloys used in aircraft engines as turbine blades. The constitutive model of Nouailhas and Freed is represented by the rheological diagram in figure 2.2.4, but all four elements in this diagram are now allowed to be anisotropic. To formulate the model, let $R : \mathcal{S}^3 \to \mathcal{S}^3$ be a linear, symmetric, positive definite mapping, which maps the subspace of matrices $\{\sigma \in \mathcal{S}^3 \mid \text{trace}(\sigma) = 0\}$ with vanishing trace into itself. Then R^{-1} exists and is also symmetric and positive definite. The mappings R and R^{-1} define norms $|\sigma|_R$ and $|\sigma|_{R^{-1}}$ on the space $\mathcal{S}^3$ by

$$|\sigma|_R = \left[(R\sigma) \cdot \sigma\right]^{1/2}, \quad |\sigma|_{R^{-1}} = \left[(R^{-1}\sigma) \cdot \sigma\right]^{1/2}.$$

With this definition the model equations take the form

$$\rho u_{tt} = \text{div}_x\, T \tag{2.2.29}$$

$$T = D(\varepsilon(\nabla_x u) - \varepsilon_p) \tag{2.2.30}$$

$$\frac{\partial}{\partial t}\varepsilon_p = \left\langle \frac{|P_0(T-Y)|_R - k}{C}\right\rangle^n \frac{RP_0(T-Y)}{|P_0(T-Y)|_R} \tag{2.2.31}$$

$$\frac{\partial}{\partial t}\varepsilon_n = \left.\left|\frac{\partial}{\partial t}\varepsilon_p\right|_{R-1}\right. \mathcal{N}^{-1}QY \tag{2.2.32}$$

with the additional equation

$$Y = \mathcal{N}(\varepsilon_p - \varepsilon_n). \tag{2.2.33}$$

Here also $\mathcal{N}, Q : \mathcal{S}^3 \to \mathcal{S}^3$ are linear, symmetric, positive definite mappings, which map $\{\sigma \in \mathcal{S}^3 \mid \mathrm{trace}(\sigma) = 0\}$ into itself. Together with R and the positive constants n, k and C they are the material parameters and must be chosen to adapt the model to the properties of the material considered. (2.2.31) and (2.2.32) are the constitutive equations for dashpots 2 and 4 in figure 2.2.4, respectively, and (2.2.33) is the constitutive equation for the Hookeian element 3. The model takes kinematic hardening into account, but not isotropic hardening. In the application of this model to study CMSX-2, a cubic single crystal superalloy, Nouailhas and Freed chose for the exponent in (2.2.31) the value $n = 4$.

Example 9. The last model we consider, proposed by L. Méric, P. Poubanne and G. Cailletaud in [145, 144], is again a model for nickel based single crystal superalloys used in gas turbine engines, but it is based on the crystallographic approach. In certain temperature ranges, for such materials inelastic deformation results principally from slips on glide planes, defined by their unit normal vectors n^s, along directions defined by the tangent unit vectors ℓ^s, $s = 1, \ldots, \nu$. The active slip planes are determined by the crystal lattice.

The constitutive model is based on this observation. For $s = 1, \ldots, \nu$ the tensor $m^s \in \mathcal{S}^3$ and the resolved shear stress $\tau^s \in \mathbb{R}$ in the sth slip plane are defined by

$$m^s = \frac{1}{2}\left(n_i^s \ell_j^s + n_j^s \ell_i^s\right)_{i,j=1,\ldots,3} \tag{2.2.34}$$

$$\tau^s = m^s \cdot T = \ell^s \cdot (Tn^s),$$

where $T \in \mathcal{S}^3$ is the macroscopic stress tensor. With the slip $\varepsilon_p^s \in \mathbb{R}$ the macroscopic plastic strain tensor $\varepsilon_p \in \mathcal{S}^3$ is defined by

$$\varepsilon_p = \sum_{s=1}^{\nu} m^s \varepsilon_p^s.$$

The constitutive equations now have to make a connection between the vectors $(\varepsilon_p^1, \ldots, \varepsilon_p^\nu)$ and $(\tau^1, \ldots, \tau^\nu)$. This is done by connecting the variables ε_p^s and τ^s for every s by equations from a constitutive model, which is similar to the models we considered in the preceding examples, and whose rheological diagram is given in figure 2.2.4, however with a nonlinear interaction between the different slip systems. The model equations are

$$\rho u_{tt} = \mathrm{div}_x T \tag{2.2.35}$$

$$T = D(\varepsilon(\nabla_x u) - \varepsilon_p) \tag{2.2.36}$$

$$\varepsilon_p = \sum_{s=1}^{\nu} m^s \varepsilon_p^s \tag{2.2.37}$$

$$\tau^s = m^s \cdot T \tag{2.2.38}$$

$$\frac{\partial}{\partial t} \varepsilon_p^s = C_1^s \left\langle |\tau^s - y^s| - k^s - \sum_{r=1}^{\nu} q_{sr} \zeta^r \right\rangle^n \frac{\tau^s - y^s}{|\tau^s - y^s|} \tag{2.2.39}$$

$$\frac{\partial}{\partial t} \varepsilon_n^s = C_2^s |\frac{\partial}{\partial t} \varepsilon_p^s| y^s \tag{2.2.40}$$

$$\frac{\partial}{\partial t} \zeta^s = C_3^s |\frac{\partial}{\partial t} \varepsilon_p^s| (C_4^s - \zeta^s) \tag{2.2.41}$$

$$y^s = \mathcal{M}^s (\varepsilon_p^s - \varepsilon_n^s). \tag{2.2.42}$$

Besides n^s and ℓ^s the constants $n, C_1^s, \ldots, C_4^s, k^s, \mathcal{M}^s > 0$, $q_{sr} \in \mathbb{R}$ are material parameters. The different slip systems interact non-linearly in equation (2.2.39) via the term $\sum_{r=1}^{\nu} q_{sr} \zeta^r$ with the interaction constants q_{sr}. The initial data must be chosen such that $0 < \zeta^s(x, 0) \le C_4^s$ for all s and x, in which case (2.2.41) implies $0 < \zeta^s(x, t) \le C_4^s$ for all s, x, and t.

In the application of this constitutive model to study a cubic nickel-based single crystal superalloy Méric, Poubanne and Cailletaud take into account 12 octahedral slip systems and 6 cubic slip systems, hence $\nu = 18$. In this case the model contains the $N = 3\nu = 54$ internal variables $\varepsilon_p^1, \ldots, \varepsilon_p^\nu, \varepsilon_n^1, \ldots, \varepsilon_n^\nu, \zeta^1, \ldots, \zeta^\nu$. For the exponent in (2.2.39) they choose $n = 3.89$.

Chapter 3

Constitutive Equations of Monotone Type and Generalized Standard Materials

In Section 3.1 we derive an energy estimate and use it to motivate our subsequently given definitions of various classes of constitutive equations. In particular, we introduce the class of constitutive equations of monotone type, which generalizes the class of constitutive equations of generalized standard materials introduced by Halphen and Nguyen Quoc Son. In Section 3.2 a uniqueness proof for dynamic initial-boundary problems with constitutive equations of monotone type and the basic ideas of the existence proof for such problems given in Chapter 4 are sketched. We shortly discuss the literature to existence proofs based on monotonicity, and also sketch the fundamental idea of existence proofs for quasi-static initial-boundary problems to some constitutive equations of monotone type, which can be found in the literature. As examples of the various classes introduced, in Section 3.3 we examine all the constitutive models presented in Section 2.2, and in Section 3.4 we finally derive a criterion, which allows us to determine whether a constitutive equation is of monotone type.

3.1 Energy Estimate and Classes of Constitutive Equations

Let (u, z) be a sufficiently smooth solution of (2.1.1) – (2.1.9). To derive an energy estimate for this solution, note that the symmetry of the stress tensor T implies

$$T \cdot \varepsilon(\nabla_x u_t) = \frac{1}{2} T \cdot (\nabla_x u_t + (\nabla_x u_t)^T) = T \cdot \nabla_x u_t$$
$$= \operatorname{div}_x (T^T u_t) - (\operatorname{div}_x T) \cdot u_t .$$

For the function $\psi = \psi(\varepsilon(\nabla_x u), z)$ we thus obtain from (2.1.1) and (2.1.8)

$$\rho(\frac{1}{2} |u_t|^2 + \psi)_t = \rho u_{tt} \cdot u_t + \rho \nabla_\varepsilon \psi \cdot \varepsilon(\nabla_x u_t) + \rho \nabla_z \psi \cdot z_t$$
$$= \rho u_{tt} \cdot u_t + T \cdot \varepsilon(\nabla_x u_t) + \rho \nabla_z \psi \cdot z_t = \operatorname{div}_x (T u_t) + \rho \nabla_z \psi \cdot z_t .$$

Integration and application of the Divergence Theorem yields the energy equality

$$\int_\Omega \rho(\tfrac{1}{2}|u_t|^2 + \psi)_t \, dx = \int_{\partial\Omega}(Tn)\cdot u_t \, dS(x) + \int_\Omega \rho\nabla_z\psi \cdot z_t \, dx\,, \qquad (3.1.1)$$

where $n = n(x)$ denotes the outward unit normal to $\partial\Omega$. Since (2.1.3) implies $z_t \in f(\varepsilon(\nabla_x u), z)$, we deduce from the dissipation inequality (2.1.9) that $\rho\nabla_z\psi\cdot z_t \le 0$. Taking the boundary conditions (2.1.5) or (2.1.6) into account, we thus obtain from (3.1.1) that

$$\frac{d}{dt}\int_\Omega \rho(\tfrac{1}{2}|u_t|^2 + \psi)dx = \int_\Omega \rho\,\nabla_z\psi\cdot z_t \, dx \le 0\,, \qquad (3.1.2)$$

which means that the sum of kinetic and free energy is decreasing. Note that in this computation we used (2.1.8), but not (2.1.2). This energy estimate is thus also valid when the relation between ε, z and T is nonlinear.

We now introduce a class of functions f, for which the dissipation inequality (2.1.9) is naturally satisfied. This class consists of those functions $f : D(f) \subseteq S^3 \times \mathbb{R}^N \to \mathbb{R}^N$ (or $f : D(f) \subseteq S^3 \times \mathbb{R}^N \to \mathcal{P}(\mathbb{R}^N)$), which can be written in the form

$$f(\varepsilon, z) = g(-\rho\nabla_z\psi(\varepsilon, z))\,, \qquad (3.1.3)$$

with a suitable free energy $\psi = D(f) \to [0, \infty)$ satisfying (2.1.8), and with a suitable monotone function $g : D(g) \subseteq \mathbb{R}^N \to \mathbb{R}^N$ (or $g : D(g) \subseteq \mathbb{R}^N \to \mathcal{P}(\mathbb{R}^N)$) satisfying

$$g(0) = 0 \quad (\text{or } 0 \in g(0))\,.$$

Remember that a vector field $g : D(g) \subseteq \mathbb{R}^N \to \mathbb{R}^N$ is called monotone, if it satisfies

$$(g(z) - g(z'))\cdot(z - z') \ge 0$$

for all $z, z' \in D(g)$. Likewise, a set-valued function (sometimes also called operator) $g : D(g) \subseteq \mathbb{R}^N \to \mathcal{P}(\mathbb{R}^N)$ is monotone, if

$$(w - w')\cdot(z - z') \ge 0$$

holds for all $z, z' \in D(g)$ and all $w \in g(z)$, $w' \in g(z')$.

To see that a function of the form (3.1.3) satisfies (2.1.9), let $(\varepsilon, z) \in D(f)$. Since $0 \in g(0)$, we obtain from the monotonicity of g that

$$\rho\nabla_z\psi(\varepsilon, z)\cdot f(\varepsilon, z) = -(-\rho\nabla_z\psi(\varepsilon, z) - 0)\cdot\Big(g(-\rho\nabla_z\psi(\varepsilon, z)) - 0\Big) \le 0\,,$$

which is (2.1.9).

For functions of the form (3.1.3) the energy estimate (3.1.2) thus holds. Below it is shown, that if such an energy inequality also holds for the difference of two solutions of (2.1.1) – (2.1.9), then we can base a uniqueness result on it. However, to derive an energy estimate for the difference of two solutions, it is not sufficient to require that f has the form (3.1.3). We need in addition that $(\varepsilon, z) \mapsto \rho\nabla\psi(\varepsilon, z) = A(\varepsilon, z)$ is a linear mapping. For the uniqueness proof we also need that the free energy ψ is non-negative and satisfies $\psi(0) = 0$. These conditions are satisfied if and only if ψ is a positive semi-definite quadratic form

$$\rho\psi(\varepsilon, z) = \frac{1}{2}[A(\varepsilon, z)]\cdot(\varepsilon, z)$$

with a symmetric, positive semi-definite matrix A. From (2.1.13) we thus obtain

$$\frac{1}{2}[A(\varepsilon, z)] \cdot (\varepsilon, z) = \rho\psi(\varepsilon, z) = \frac{1}{2}[D(\varepsilon - Bz)] \cdot (\varepsilon - Bz) + \psi_1(z),$$

from which we easily conclude that ψ must necessarily be of the form

$$\rho\psi(\varepsilon, z) = \frac{1}{2}[D(\varepsilon - Bz)] \cdot (\varepsilon - Bz) + \frac{1}{2}(Lz) \cdot z$$

with a symmetric $N \times N$–matrix L. Since $B : \mathbb{R}^N \to S^3$, for every $z \in \mathbb{R}^N$ we can choose an element $\varepsilon \in S^3$ with $\varepsilon - Bz = 0$. Because ψ must be non-negative, we obtain for this ε that

$$\frac{1}{2}(Lz) \cdot z = \frac{1}{2}[D(\varepsilon - Bz)] \cdot (\varepsilon - Bz) + \frac{1}{2}(Lz) \cdot z = \rho\psi(\varepsilon, z) \geq 0,$$

which shows that L must be a positive semi-definite matrix. Differentiating the above expression for $\rho\psi$ yields

$$- \rho\nabla_z\psi(\varepsilon, z) = B^T D(\varepsilon - Bz) - Lz = \overline{M}\varepsilon - Mz \qquad (3.1.4)$$

with linear mappings

$$\overline{M} = B^T D : S^3 \to \mathbb{R}^N,$$

and

$$M = B^T D B + L : \mathbb{R}^N \to \mathbb{R}^N.$$

Since $D : S^3 \to S^3$ is symmetric and positive definite, it follows that the mapping M is symmetric and positive semi-definite, and thus is represented by a symmetric, positive semi-definite $N \times N$–matrix. In this work we require that M is not only positive semi-definite, but positive definite. In almost all applications of the theory this requirement is satisfied. This is well illustrated by the discussion of the constitutive model of Méric, Poubanne and Cailletaud following below in Section 3.3.

These considerations motivate our definition of the class of constitutive equations of monotone type, which we give now. In this definition we need the notions of a proper function and of the subdifferential of a proper function: a function $\chi : D(\chi) \subseteq \mathbb{R}^N \to (-\infty, \infty]$ is called proper, if there exists $z \in D(\chi)$ with $\chi(z) < \infty$. The subdifferential $\partial\chi : D(\chi) \subseteq \mathbb{R}^N \to \mathcal{P}(\mathbb{R}^N)$ of a proper function χ is defined by

$$w \in \partial\chi(z) \iff \forall \bar{z} \in D(\chi) : \chi(\bar{z}) \geq \chi(z) + w \cdot (\bar{z} - z).$$

Definition 3.1.1 *We say that a pair (f, B) of a function*

$$f : D(f) \subseteq S^3 \times \mathbb{R}^N \to \mathbb{R}^N \quad (f : D(f) \subseteq S^3 \times \mathbb{R}^N \to \mathcal{P}(\mathbb{R}^N))$$

and a linear mapping $B : \mathbb{R}^N \to S^3$ is of pre-monotone type, if there exists a symmetric, positive definite $N \times N$–matrix M with the property that the symmetric $N \times N$–matrix

$$L = M - B^T D B \qquad (3.1.5)$$

is positive semi-definite, and if there exists a function

$$g : D(g) \subseteq \mathbb{R}^N \to \mathbb{R}^N \quad (g : D(g) \subseteq \mathbb{R}^N \to \mathcal{P}(\mathbb{R}^N))$$

such that

$$f(\varepsilon, z) = g(-\rho \nabla_z \psi(\varepsilon, z)) = g(B^T D\varepsilon - Mz)$$

for all $(\varepsilon, z) \in D(f)$, with the positive semi-definite quadratic form

$$\rho\psi(\varepsilon, z) = \frac{1}{2}[D(\varepsilon - Bz)] \cdot (\varepsilon - Bz) + \frac{1}{2}(Lz) \cdot z.$$

If the function g is the gradient (or subdifferential)

$$g = \nabla\chi \quad (g = \partial\chi)$$

of a proper function $\chi : D(\chi) \subseteq \mathbb{R}^N \to (-\infty, \infty]$, then we say that the pair (f, B) is of gradient type. If the function g is monotone and satisfies

$$g(z) \cdot z \geq 0 \quad (w \cdot z \geq 0), \tag{3.1.6}$$

for all $z \in D(g)$ (and all $w \in g(z)$), then we say that the pair (f, B) is of monotone type. If the function g is the gradient (or subdifferential) of a proper convex function $\chi : D(\chi) \subseteq \mathbb{R}^N \to (-\infty, \infty]$ and satisfies (3.1.6), then we say that the pair (f, B) is of monotone-gradient type or of the type of generalized standard materials. By $\mathcal{M}^*, \mathcal{G}, \mathcal{M}, \mathcal{MG}$ we denote the sets of pairs (f, B), which are of pre-monotone, of gradient, of monotone, or of monotone-gradient type, respectively. Accordingly, we say that the constitutive equations (2.1.2), (2.1.3) are of pre-monotone, gradient, monotone, or monotone-gradient type, if the pair (f, B) from the constitutive equations is of the respective type. Then the constitutive equations can be written in the form

$$T = D(\varepsilon - Bz)$$

$$z_t = f(\varepsilon, z) = g(-\rho \nabla_z \psi(\varepsilon, z))$$

$$\left(z_t \in f(\varepsilon, z) = g(-\rho \nabla_z \psi(\varepsilon, z))\right).$$

The function ψ (which is not necessarily uniquely determined by these conditions) is called free energy associated to f.

Remark For monotone g, condition (3.1.6) is satisfied if

$$0 \in g(0), \tag{3.1.7}$$

since then for $z \in D(g)$ and $w \in g(z)$

$$(w - 0) \cdot (z - 0) = w \cdot z \geq 0.$$

At the same time this inequality shows that if (3.1.6) holds, but $0 \notin g(0)$, then we can always extend g to a larger monotone operator $\bar{g} : \mathcal{D}(\bar{g}) \subseteq \mathbb{R}^N \to \mathcal{P}(\mathbb{R}^N)$ by defining $\bar{g}(z) = g(z)$ if $z \neq 0$ and $\bar{g}(0) = g(0) \cup \{0\}$. Here we set $g(0) = \emptyset$ if $0 \notin D(g)$. Since by definition a maximal monotone operator does not have a proper monotone extension, conditions (3.1.6) and (3.1.7) are therefore equivalent for maximal monotone operators.

The class of constitutive equations of generalized standard materials (matériaux standards généralisés) was introduced by Halphen and Nguyen Quoc Son in [88]. Since gradients or subdifferentials of proper convex functions are monotone, the class $\mathcal{MG}$ is a subclass of $\mathcal{M} \cap \mathcal{G}$.

3.2   Uniqueness and Existence for Dynamic and Quasi-Static Problems: Basic Ideas of the Proofs and Results in the Literature

For constitutive equations of monotone type the energy estimate (3.1.2) can be generalized to prove uniqueness of the solution. Namely, the linearity of $\nabla\psi$ allows us to derive an energy estimate for the difference of two smooth solutions (u, z) and $(\overline{u}, \overline{z})$ of (2.1.1) – (2.1.9) by a computation completely analogous to the one that led to (3.1.2). The result is

$$\frac{d}{dt} \int_\Omega \rho[\frac{1}{2} |u_t - \overline{u}_t|^2 + \psi(\varepsilon(\nabla_x(u - \overline{u})), z - \overline{z})]\, dx \qquad (3.2.1)$$

$$= - \int_\Omega [-\rho\nabla_z\psi(\varepsilon, z) + \rho\nabla_z\psi(\overline{\varepsilon}, \overline{z})] \cdot [g(-\rho\nabla_z\psi(\varepsilon, z)) - g(-\rho\nabla_z\psi(\overline{\varepsilon}, \overline{z}))]\, dx \le 0\,,$$

where we also used (3.1.3). The term in the middle is less or equal to zero because of the monotonicity of g. From this inequality we immediately obtain that the initial-boundary value problem (2.1.1) – (2.1.9) has at most one smooth solution, since (3.2.1) implies

$$\int_\Omega \rho[\frac{1}{2} |(u_t - \overline{u}_t)(x, t)|^2 + \psi\big(\varepsilon(\nabla_x(u - \overline{u})(x, t)), (z - \overline{z})(x, t)\big)]\, dx \qquad (3.2.2)$$

$$\le \int_\Omega \rho\left[\frac{1}{2} |(u_t - \overline{u}_t)(x, 0)|^2 + \psi\big(\varepsilon(\nabla_x(u - \overline{u})(x, 0)), (z - \overline{z})(x, 0)\big)\right]\, dx = 0\,,$$

if $u(x, 0) = \overline{u}(x, 0)$, $u_t(x, 0) = \overline{u}_t(x, 0)$, $z(x, 0) = \overline{z}(x, 0)$. Since by assumption ψ is positive semi-definite, we obtain

$$\int_\Omega |(u_t - \overline{u}_t)(x, t)|^2 dx = 0\,,$$

hence $u_t(x, t) = \overline{u}_t(x, t)$ and therefore $u(x, t) = \overline{u}(x, t)$. Using (3.1.5) we then conclude from (3.2.2) that

$$0 = \int_\Omega \rho\psi(0, (z - \overline{z})(x, t))\, dx$$

$$= \int_\Omega \frac{1}{2}[DB(z - \overline{z})] \cdot B(z - \overline{z}) + \frac{1}{2}[L(z - \overline{z})] \cdot (z - \overline{z})\, dx$$

$$= \int_\Omega \frac{1}{2}[(B^T DB + L)(z - \overline{z})] \cdot (z - \overline{z})\, dx = \int_\Omega \frac{1}{2}[M(z - \overline{z})] \cdot (z - \overline{z})\, dx\,,$$

which implies that $z(x, t) = \overline{z}(x, t)$, since by assumption M is positive definite. Consequently, the solution of the initial-boundary value problem is unique.

It is suggested by these considerations that for (f, B) of monotone type, not only uniqueness, but also existence of a solution can be proved using the theory of evolution equations for monotone operators. This is done in the next chapter under the stronger hypotheses that the associated free energy ψ is not only positive semi-definite, but positive definite, and that g is not only monotone, but maximal monotone.

To prove this result we show that the solution of the initial-boundary value problem can be reduced to the solution of an initial value problem for the equation

$$w_t + Cw = 0,$$

with

$$(x,t) \mapsto w(x,t) = (u_t(x,t), \varepsilon(x,t), z(x,t)) : \Omega \times [0,\infty) \to W = \mathbb{R}^3 \times \mathcal{S}^3 \times \mathbb{R}^N$$

and with an operator

$$C : D(C) \subseteq L^2(\Omega, W) \to \mathcal{P}(L^2(\Omega, W)),$$

which is maximal monotone if the bilinear form

$$\langle (v, \varepsilon, z), (\overline{v}, \overline{\varepsilon}, \overline{z}) \rangle = \int_\Omega \rho v(x) \cdot \overline{v}(x) + a((\varepsilon, z)(x), (\overline{\varepsilon}, \overline{z})(x)) \, dx$$

with

$$a((\varepsilon, z), (\overline{\varepsilon}, \overline{z})) = [D(\varepsilon - Bz)] \cdot (\overline{\varepsilon} - B\overline{z}) + (Lz) \cdot \overline{z}$$

defines a scalar product on the space $L^2(\Omega, W)$, and if this space is equipped with this scalar product. The bilinear form $\langle (v, \varepsilon, z), (\overline{v}, \overline{\varepsilon}, \overline{z}) \rangle$ is a scalar product if and only if the bilinear form $a((\varepsilon, z), (\overline{\varepsilon}, \overline{z}))$ is a scalar product on the space $\mathcal{S}^3 \times \mathbb{R}^N$. This is the case if and only if ψ is positive definite, since

$$\rho\psi(\varepsilon, z) = \frac{1}{2} a((\varepsilon, z), (\varepsilon, z)).$$

Here the hypothesis that ψ is positive definite enters the proof.

This existence result applies in particular to initial-boundary value problems with constitutive relations of the class $\mathcal{MG}$, the class of constitutive relations of generalized standard materials, and implies that a unique solution exists if in

$$T = D(\varepsilon - Bz)$$

$$z_t = \nabla\chi(-\rho\nabla_z\psi(\varepsilon, z)) \quad \text{or} \quad z_t \in \partial\chi(-\rho\nabla_z\psi(\varepsilon, z)), \tag{3.2.3}$$

the free energy ψ is positive definite and the gradient $\nabla\chi$ or the subdifferential $\partial\chi$ is maximal monotone. This last condition is satisfied, for example, if the proper convex function χ with values in $(-\infty, \infty]$ is defined on all of $\mathbb{R}^N$.

The majority of existence results which can be found in the mathematical literature to initial-boundary value problems with internal variables concern constitutive relations of monotone-gradient type. For example, a closer examination shows that the constitutive relations studied in [16, 17, 41, 62, 88, 102, 107, 108, 133, 146, 156, 165, 166, 201, 206] are of monotone-gradient type. These existence results can be roughly divided into two groups. The first group, which is small, consists of existence results for initial-boundary value problems of the form (2.1.1) – (2.1.9) with special and more or less restricted constitutive equations of monotone-gradient type, for which the associated free energy is positive definite. Such existence results are given in [17, 133], where inhomogeneous boundary conditions and an additional inhomogeneous term on the right hand side of (2.1.1) are also allowed. The existence theorem from the following chapter and its method of proof are generalizations of these existence results and proofs for the class of constitutive relations of monotone type, for which the vector field g from Definition 3.1.1 is maximal monotone, and whose associated free energy is positive definite. From the mathematical point of view, in this generality the existence result is rather satisfactory, but from the point of view of the applications and of the transformation theory developed in this work, the positivity condition is unfortunate, since even after transformation most constitutive relations do not satisfy it. This is shown by the examples studied below and by the transformation theory developed later.

This positivity condition is not needed and positive semi-definite free energy suffices in the second group. Also this group, which is larger, consists of existence results to special and restricted constitutive relations of monotone-gradient type, but in this group the initial-boundary value problems considered are quasi-static. The initial-boundary value problem is called quasi-static if the equation (2.1.1) is replaced by

$$- \operatorname{div}_x T(x, t) = h(x, t) \tag{3.2.4}$$

for a given function $h : \Omega \times [0, \infty) \to \mathbb{R}^3$. The proofs of these existence results are based on a reduction of the problem which can be made in the quasi-static case. For simplicity, we explain this reduction only for the initial-boundary value problem with homogeneous Dirichlet boundary condition and with constitutive relations from Examples 1 or 2 of Section 2.2, the Prandtl-Reuss or Maxwell models.

In the quasi-static case the system of partial and ordinary differential equations consists of (3.2.4) and of

$$T = D(\varepsilon(\nabla_x u) - \varepsilon_p) \tag{3.2.5}$$

$$\frac{\partial}{\partial t} \varepsilon_p \in g_K(T), \tag{3.2.6}$$

and the boundary condition is

$$u(x, t) = 0, \quad x \in \partial\Omega. \tag{3.2.7}$$

The function T is sought in the space $L^{2,\text{loc}}\big([0, \infty), L^2(\Omega, S^3)\big)$, and it is assumed that a function $\tau \in L^{2,\text{loc}}([0, \infty), L^2(\Omega, S^3))$ is known satisfying

$$-\operatorname{div}_x \tau = h.$$

Such a function can often be determined easily. In the case of rate independent problems it is connected with the *collapse* and *shakedown theorems* from mechanics, cf. [198] or [15, p. 156ff], for example. The difference $\sigma = T - \tau$ fulfills

$$\operatorname{div}_x \sigma = 0,$$

and thus for every t the function $\sigma(t)$ belongs to the closed subspace

$$\mathcal{D}_0 = \{w \in L^2(\Omega, S^3) \mid \operatorname{div} w = 0\}$$

of $L^2(\Omega, S^3)$. Here $\sigma(t)$ denotes the function $x \mapsto \sigma(x, t)$. On the other hand, the boundary condition (3.2.7) yields for the symmetric gradient $\varepsilon(\nabla_x u)$ and for $w \in \mathcal{D}_0$ that

$$\int_\Omega \varepsilon\big(\nabla_x u(x, t)\big) \cdot w(x, t) \, dx = -\int_\Omega u(x, t) \cdot \operatorname{div}_x w(x, t) \, dx = 0,$$

which means that $\varepsilon\big(\nabla_x u(t)\big)$ is orthogonal to $\mathcal{D}_0$. For the orthogonal projector Q of $L^2(\Omega, \mathcal{S}^3)$ onto $\mathcal{D}_0$ we thus obtain that $Q\varepsilon = Q\varepsilon_t = 0$.

We now insert (3.2.5) into (3.2.6) to eliminate ε_p. The result is

$$\varepsilon_t - D^{-1}T_t \in g_K(T).$$

If we apply Q to both sides of this equation and write $T = \sigma + \tau$, we obtain

$$QD^{-1}\sigma_t \in -Qg_K(\sigma + \tau) - QD^{-1}\tau_t$$

or

$$\sigma_t \in -(QD^{-1})^{-1}Qg_K(\sigma + \tau) - (QD^{-1})^{-1}QD^{-1}\tau_t, \qquad (3.2.8)$$

with QD^{-1} defined by

$$[QD^{-1}w](x) = [Q(D^{-1}w(\cdot))](x).$$

In this equation $(QD^{-1})^{-1}QD^{-1}$ cannot be replaced by the identity on $\mathcal{D}_0$, since $(QD^{-1})^{-1}$ denotes the inverse of

$$QD^{-1}\big|_{\mathcal{D}_0} : \mathcal{D}_0 \to \mathcal{D}_0,$$

whereas $\tau_t(t)$ is not necessarily contained in $\mathcal{D}_0$. Note that this inverse exists, since for $\sigma, \tau \in \mathcal{D}_0$

$$\int_\Omega [Q(D^{-1}\sigma(x))] \cdot \tau(x)\,dx = \int_\Omega [D^{-1}\sigma(x)] \cdot \tau(x)\,dx = \int_\Omega \sigma(x) \cdot [Q(D^{-1}\tau(x))]\,dx,$$

which implies that $QD^{-1}\big|_{\mathcal{D}_0} = QD^{-1}Q\big|_{\mathcal{D}_0}$ is symmetric and positive definite. (3.2.8) can be viewed as evolution equation for $\sigma \in L^{2,\mathrm{loc}}([0, \infty), \mathcal{D}_0)$ containing a family of operators

$$w \mapsto (QD^{-1})^{-1}Qg_K(w + \tau(t)) : \mathcal{D}_0 \to \mathcal{P}(\mathcal{D}_0),$$

which depend on the parameter t, and which in the scalar product

$$(w, v) = \int_\Omega [QD^{-1}w](x) \cdot v(x)\,dx$$

on $\mathcal{D}_0$ are monotone for every t if $g_K : \mathcal{S}^3 \to \mathcal{P}(\mathcal{S}^3)$ is a monotone operator. This is shown by the following computation, where we omit any technical details and assume that $g_K : \mathcal{S}^3 \to \mathcal{S}^3$ is single-valued. Since Q is a projector onto $\mathcal{D}_0$ orthogonal with respect to the usual scalar product of $L^2(\Omega)$, we obtain for $v, w \in \mathcal{D}_0$ that

$$\Big((QD^{-1})^{-1}Q[g_K(v + \tau) - g_K(w + \tau)], v - w\Big)$$

$$= \int_\Omega Q[g_K(v(\cdot) + \tau(\cdot, t)) - g_K(w(\cdot) + \tau(\cdot, t))](x) \cdot (v(x) - w(x))\,dx$$

$$= \int_\Omega \Big(g_K(v(x) + \tau(x, t)) - g_K(w(x) + \tau(x, t))\Big)$$

$$\cdot \Big((v(x) + \tau(x, t)) - (w(x) + \tau(x, t))\Big)\,dx \geq 0.$$

Therefore the operators are monotone. The determination of the stress T is thus reduced to the solution of equation (3.2.8) for σ under suitable initial conditions. This is an evolution equation for a one-parameter family of operators which are monotone, even if the free energy is not positive definite.

Proofs of existence and regularity of solutions to quasi-static initial-boundary value problems with special constitutive equations of monotone-gradient type are contained in [16, 62, 107, 108, 133, 156, 201, 206], for example. We surmise that these proofs can be generalized to all constitutive equations of monotone type, for which the vector field g from Definition 3.1.1 is maximal monontone, and which satisfy some additional, weak conditions. Nevertheless, in this work we do not attempt to do this, since our main concern is to develop the transformation theory for constitutive equations. This transformation theory is of equal importance for dynamic problems, that is, for systems of equations of the form (2.1.1) – (2.1.3), and for quasi-static problems.

An existence theory of equal generality as in the two groups just described has not been found for the dynamic initial-boundary value problem (2.1.1) – (2.1.9) with constitutive equations of monotone type, whose associated free energy is not positive definite, but only positive semi-definite. Some results are known, for example, the existence theorem in the book [62] for the dynamic problem for the Prandtl-Reuss law and the result in [41] fall in this category, but to develop a general existence theory seems to be an open problem.

3.3 Examples of Constitutive Equations Revisited

To illustrate the notions introduced in Definition 3.1.1, we show in this section that all the examples of constitutive equations presented in Section 2.2 are of pre-monotone type, and for some of these examples we also show that they are of monotone-gradient type.

At the end of this chapter criteria are derived, which guarantee that constitutive equations are of monotone type, and a subset of these criteria already guarantees, that the equations belong to the class of pre-monotone type. These criteria could be used in the following investigations to show that the constitutive equations are of pre-monotone type; but since it is more instructive, instead of using these criteria we prove for every one of the examples that the constitutive equations are of pre-monotone type by explicitly writing them in pre-monotone form.

The model of Prandtl-Reuss. In this constitutive model the vector of internal variables is $z(x,t) = \varepsilon_p(x,t) \in \mathcal{S}^3$. As we noted in Section 2.2, in the discussion of this and the following examples it is convenient to identify $\mathcal{S}^3$ with $\mathbb{R}^6$. If $\{\sigma^{(1)}, \ldots, \sigma^{(6)}\}$ is an orthonormal basis of $\mathcal{S}^3$, then an isomorphism between $\mathbb{R}^6$ and $\mathcal{S}^3$ is given by

$$w = (w_1, \ldots, w_6) \mapsto w_1\sigma^{(1)} + \ldots + w_6\,\sigma^{(6)} : \mathbb{R}^6 \to \mathcal{S}^3 \subseteq \mathbb{R}^9, \qquad (3.3.1)$$

a coordinate mapping of the six-dimensional manifold $\mathcal{S}^3$.

With this identification we set $N = 6$, define a positive semi-definite quadratic form $\rho\psi$ on $\mathcal{S}^3 \times \mathbb{R}^N$ by

$$\rho\psi(\varepsilon, \varepsilon_p) = \frac{1}{2}\left[D(\varepsilon - \varepsilon_p)\right] \cdot (\varepsilon - \varepsilon_p), \tag{3.3.2}$$

and substitute $-\rho\nabla_{\varepsilon_p}\psi(\varepsilon, \varepsilon_p) = D(\varepsilon - \varepsilon_p)$ for T in (2.2.2). Moreover, if we define the characteristic function $\chi_K : \mathcal{S}^3 \to [0, \infty]$ of the closed convex set $K \subseteq \mathcal{S}^3$ by

$$\chi_K(z) = \begin{cases} 0 \,, & \text{if } z \in K \\ \infty \,, & \text{if } z \in \mathcal{S}^3 \backslash K \,, \end{cases}$$

then it is easily seen that the mapping $g_K : \mathcal{S}^3 \to P(\mathcal{S}^3)$ defined in (2.2.3) is equal to the subdifferential $\partial\chi_K : \mathcal{S}^3 \to \mathcal{P}(\mathcal{S}^3)$ of χ_K. Therefore for the Prandtl-Reuss model the equations (2.1.1) – (2.1.3) obtain the form

$$\rho u_{tt} = \operatorname{div}_x T$$
$$T = D(\varepsilon(\nabla_x u) - z) \tag{3.3.3}$$
$$z_t \in \partial\chi_K\left(-\rho\nabla_z\,\psi(\varepsilon(\nabla_x u), z)\right) \tag{3.3.4}$$

with $z = \varepsilon_p$. In this case the mapping B in (2.1.2) is the identity on $\mathcal{S}^3$.

Since for the quadratic form (3.3.2) the linear, symmetric mapping M defined by $Mz = \rho\nabla_z\psi(0, z)$ satisfies

$$M = B^T D B = D : \mathcal{S}^3 \to \mathcal{S}^3 \,,$$

and thus is positive definite, it follows immediately that the hypotheses of Definition 3.1.1 are satisfied, and that (3.3.3), (3.3.4) are constitutive relations of gradient type. Moreover, if the convex set K contains 0, then $0 \in \partial\chi_K(0)$, by definition of the subdifferential. As we remarked after Definition 3.1.1, this implies that (3.1.6) is satisfied, and since χ_K is a convex function, if follows that (3.3.3), (3.3.4) are of monotone-gradient type and consequently satisfy the dissipation inequality (2.1.9). Therefore the constitutive equations of Prandtl-Reuss are thermodynamically admissible if $0 \in K$.

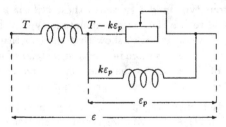

Figure 3.3.1

The quadratic form (3.3.2), the associated free energy for the constitutive equations of Prandtl-Reuss, is positive semi-definite, but not positive definite. Therefore the Prandtl-Reuss model furnishes an example for a constitutive model, whose associated free energy is positive semi-definite, but not positive definite. It is instructive to study how the rheological diagram for the Prandtl-Reuss model displayed in figure 2.2.1 must be changed to obtain a constitutive model with positive definite free energy. This is achieved in figure 3.3.1 by adding a linear spring with spring constant $k > 0$. The constitutive model corresponding to this diagram is obtained by replacing the relation (2.2.2) by the new relation

$$\frac{\partial}{\partial t}\varepsilon_p \in g_K(T - k\varepsilon_p) = \partial\chi_K(T - k\varepsilon_p),$$

which amounts to replacing (3.3.4) by

$$z_t \in \partial\chi\left(-\rho\nabla_z\tilde{\psi}(\varepsilon(\nabla_x u), z)\right),$$

where the free energy

$$\rho\tilde{\psi}(\varepsilon, z) = \rho\psi(\varepsilon, z) + \frac{1}{2}kz \cdot z = \frac{1}{2}[D(\varepsilon - z)] \cdot (\varepsilon - z) + \frac{k}{2}|z|^2$$

is now positive definite.

The model of Maxwell. In exactly the same way it follows that this model is also of pre-monotone type: Define ψ as in (3.3.2), substitute $-\rho\nabla_{\varepsilon_p}\psi(\varepsilon, \varepsilon_p) = D(\varepsilon - \varepsilon_p)$ for T in (2.2.4), and set $z = \varepsilon_p$. Then for the Maxwell model the equations (2.1.1) – (2.1.3) become

$$\rho u_{tt} = \text{div}_x T$$

$$T = D(\varepsilon - z) \tag{3.3.5}$$

$$z_t = g(-\rho\nabla_z\psi(\varepsilon, z)), \tag{3.3.6}$$

where the constitutive equations (3.3.5), (3.3.6) are of pre-monotone type. If g is of the form

$$g(\sigma) = \Gamma(|P_0\sigma|)\frac{P_0\sigma}{|P_0\sigma|}$$

with a continuous, monotonically increasing function $\Gamma : [0, \infty) \to [0, \infty)$ satisfying $\Gamma(0) = 0$, then every primitive $\tilde{\Gamma} : [0, \infty) \to \mathbb{R}$ of Γ is convex, and therefore also the function

$$\sigma \mapsto \chi(\sigma) = \tilde{\Gamma}(|P_0\sigma|) : \mathcal{S}^3 \to \mathbb{R}.$$

Using that $\sigma \mapsto P_0\sigma = \sigma - \frac{1}{3}\text{trace}(\sigma)I : \mathcal{S}^3 \to \mathcal{S}^3$ is an orthogonal projection, and hence $P_0^T = P_0$, we obtain

$$\nabla\chi(\sigma) = \Gamma(|P_0\sigma|)\frac{P_0\sigma}{|P_0\sigma|} = g(\sigma).$$

Therefore (3.3.6) can be written as

$$z_t = \nabla\chi(-\rho\nabla_z\psi(\varepsilon, z)),$$

with $\nabla\chi(0) = \lim_{\sigma \to 0}\Gamma(|P_0\sigma|)\frac{P_0\sigma}{|P_0\sigma|} = 0$. This shows that under these hypotheses for g the constitutive equations are of monotone-gradient type and consequently thermodynamically admissible.

Models obtained by combination of Hookeian elements and dashpots. The constitutive equations (2.2.9) –(2.2.11) with rheological diagram given by figure

2.2.4 are also of pre-monotone type. To show this we set $N = 12$, choose for the vector of internal variables $z = (\varepsilon_p, \varepsilon_n) \in S^3 \times S^3 \cong \mathbb{R}^N$, and define the mapping $B : S^3 \times S^3 \to S^3$ by

$$Bz = B(\hat{z}, \tilde{z}) = \hat{z} \in S^3.$$

The transpose $B^T : S^3 \to S^3 \times S^3$ is given by $B^T \hat{z} = (\hat{z}, 0)$, hence $B^T DBz = B^T DB(\hat{z}, \tilde{z}) = (D\hat{z}, 0)$ and $B^T D\varepsilon = (D\varepsilon, 0)$. With the constant $\mathcal{M} > 0$ from (2.2.10), (2.2.11) the linear symmetric mapping $M : S^3 \times S^3 \to S^3 \times S^3$ is defined by

$$Mz = M(\hat{z}, \tilde{z}) = \left(D\hat{z} + \mathcal{M}(\hat{z} - \tilde{z}), -\mathcal{M}(\hat{z} - \tilde{z}) \right).$$

This mapping is positive definite and

$$L = M - B^T DB$$

is positive semi-definite because D is by assumption a symmetric positive definite mapping on S^3. This implies

$$(Mz) \cdot z = [M(\hat{z}, \tilde{z})] \cdot (\hat{z}, \tilde{z}) = (D\hat{z}) \cdot \hat{z} + \mathcal{M}|\hat{z} - \tilde{z}|^2 > 0$$

for $z \neq 0$, and

$$(Lz) \cdot z = (Mz) \cdot z - (D\hat{z}, 0) \cdot (\hat{z}, \tilde{z}) = \mathcal{M}|\hat{z} - \tilde{z}|^2 \geq 0.$$

Finally, if we define $g : S^3 \times S^3 \to S^3 \times S^3$ by

$$g(z) = g(\hat{z}, \tilde{z}) = \left(\hat{g}(\hat{z}), \tilde{g}(\tilde{z}) \right) \tag{3.3.7}$$

for the mappings $\hat{g}$ and $\tilde{g}$ from (2.2.10), (2.2.11), then a short computation shows that the equations (2.2.8) – (2.2.11) can be written in the form

$$\rho u_{tt} = \operatorname{div}_x T$$
$$T = D(\varepsilon - Bz)$$
$$z_t = g(B^T D\varepsilon - Mz) = g(-\rho \nabla_z \psi(\varepsilon, z))$$

where

$$\rho \psi(\varepsilon, z) = \frac{1}{2}[D(\varepsilon - Bz)] \cdot (\varepsilon - Bz) + \frac{1}{2}(Lz) \cdot z$$

$$= \frac{1}{2}[D(\varepsilon - \varepsilon_p)] \cdot (\varepsilon - \varepsilon_p) + \frac{\mathcal{M}}{2}|\varepsilon_p - \varepsilon_n|^2.$$

This shows that the equations (2.2.9) – (2.2.11) are of pre-monotone type. Moreover, if the functions $\hat{g}, \tilde{g}$ satisfy $\hat{g}(0) = \tilde{g}(0) = 0$ and are gradients of convex functions:

$$\hat{g}(\hat{z}) = \nabla \hat{\chi}(\hat{z}), \quad \tilde{g}(\tilde{z}) = \nabla \tilde{\chi}(\tilde{z}),$$

then g defined in (3.3.7) satisfies $g(0) = 0$ and is the gradient field of the convex function $\chi(z) = \hat{\chi}(\hat{z}) + \tilde{\chi}(\tilde{z})$:

$$g(z) = g(\hat{z}, \tilde{z}) = \left(\nabla \hat{\chi}(\hat{z}), \nabla \tilde{\chi}(\tilde{z}) \right) = \nabla \chi(z).$$

Therefore under these assumptions the constitutive equations (2.2.9) – (2.2.11) are of monotone-gradient type and thus thermodynamically admissible. In particular, these assumptions are satisfied if $\hat{g}, \tilde{g}$ are equal to the functions on the right hand sides of (2.2.6) and (2.2.7), provided that $\hat{\Gamma}$ and $\tilde{\Gamma}$ are monotonically increasing.

Models of Chaboche and Bodner-Partom. It can be seen by adaption of the preceding definitions that the constitutive model of Chaboche is of pre-monotone type. We set $N = 13$ and choose $z = (\varepsilon_p, \varepsilon_n, \zeta) \in \mathcal{S}^3 \times \mathcal{S}^3 \times \mathbb{R} \cong \mathbb{R}^N$ to be the vector of internal variables. The mapping $B : \mathcal{S}^3 \times \mathcal{S}^3 \times \mathbb{R} \to \mathcal{S}^3$ is now defined by

$$Bz = B(\hat{z}, \tilde{z}, \eta) = \hat{z} \in \mathcal{S}^3, \tag{3.3.8}$$

with transpose $B^T : \mathcal{S}^3 \to \mathcal{S}^3 \times \mathcal{S}^3 \times \mathbb{R}$ given by $B^T \hat{z} = (\hat{z}, 0, 0)$. Hence $B^T DBz = (D\hat{z}, 0, 0)$ and $B^T D\varepsilon = (D\varepsilon, 0, 0)$. The mapping $M : \mathcal{S}^3 \times \mathcal{S}^3 \times \mathbb{R} \to \mathcal{S}^3 \times \mathcal{S}^3 \times \mathbb{R}$ is defined by

$$Mz = M(\hat{z}, \tilde{z}, \eta) = \big(D\hat{z} + \mathcal{M}(\hat{z} - \tilde{z}), -\mathcal{M}(\hat{z} - \tilde{z}), \eta\big), \tag{3.3.9}$$

where $\mathcal{M} > 0$ is the constant from (2.2.17). This mapping is symmetric, positive definite, and $L = M - B^T DB$ is positive semi-definite, because

$$(Mz) \cdot z = [M(\hat{z}, \tilde{z}, \eta)] \cdot (\hat{z}, \tilde{z}, \eta) = (D\hat{z}) \cdot \hat{z} + \mathcal{M}|\hat{z} - \tilde{z}|^2 + \eta^2 > 0$$

for $z \neq 0$, and

$$(Lz) \cdot z = (Mz) \cdot z - (D\hat{z}, 0, 0) \cdot (\hat{z}, \tilde{z}, \eta) = \mathcal{M}|\hat{z} - \tilde{z}|^2 + \eta^2 \geq 0.$$

Finally, if we define $g : \mathcal{S}^3 \times \mathcal{S}^3 \times (0, \zeta_0] \to \mathcal{S}^3 \times \mathcal{S}^3 \times \mathbb{R}$ by

$$g(z) = g(\hat{z}, \tilde{z}, \eta) = \big(\hat{g}(\hat{z}, \eta), \tilde{g}(\hat{z}, \tilde{z}, \eta), \gamma(\hat{z}, \eta)\big) \tag{3.3.10}$$

with

$$\hat{g}(\hat{z}, \eta) = \Big\langle \frac{|P_0 \hat{z}| + \eta}{C_1} \Big\rangle^n \frac{P_0 \hat{z}}{|P_0 \hat{z}|}$$

$$\tilde{g}(\hat{z}, \tilde{z}, \eta) = C_2 |\tilde{z}|^m \frac{\tilde{z}}{|\tilde{z}|} + C_3 \Big\langle \frac{|P_0 \hat{z}| + \eta}{C_1} \Big\rangle^n \tilde{z}$$

$$\gamma(\hat{z}, \eta) = C_4 \Big\langle \frac{|P_0 \hat{z}| + \eta}{C_1} \Big\rangle^n (\zeta_0 + \eta),$$

then it can be immediately verified that equations (2.2.12) – (2.2.17) can be written in the form

$$\rho u_{tt} = \text{div}_x T$$
$$T = D(\varepsilon - Bz)$$
$$z_t = g(B^T D\varepsilon - Mz) = g(-\rho \nabla_z \psi(\varepsilon, z)),$$

where

$$\rho \psi(\varepsilon, z) = \frac{1}{2}[D(\varepsilon - Bz)] \cdot (\varepsilon - Bz) + \frac{1}{2}(Lz) \cdot z =$$

$$= \frac{1}{2}[D(\varepsilon - \varepsilon_p)] \cdot (\varepsilon - \varepsilon_p) + \frac{\mathcal{M}}{2}|\varepsilon_p - \varepsilon_n|^2 + \frac{1}{2}\zeta^2.$$

To obtain the last equality we set $z = (\varepsilon_p, \varepsilon_n, \zeta)$, and thus return to the original interior variables. This proves that the constitutive model of Chaboche is of pre-monotone type.

It is considerably more difficult to determine under what conditions the constitutive equations of Chaboche are of monotone or monotone-gradient type. A convex potential of the function g cannot be given in such an obvious and simple way as in the preceding examples. This is a common difficulty for models from the engineering literature, and also arises for the remaining examples presented in Section 2.2. In Section 3.4 we therefore give criteria, which allow us to decide, whether constitutive equations are of monotone type. The application of these criteria is demonstrated in Chapter 8, where the model of Bodner and Partom is investigated in detail. Since it is shown there that this model is of pre-monotone type, we do not discuss the constitutive equations of Bodner and Partom here. However, since it is of basic interest for the mathematical investigations in this work, we mention already here that it will turn out that the model of Bodner and Partom is neither of monotone nor of gradient type, and that this also holds for the model of Chaboche and for all remaining models given in Section 2.2.

Models of Hart and Miller. Exactly as for the model of Chaboche it can be shown that the constitutive equations of Hart and of Miller are of pre-monotone type. The mappings B and M are defined as in (3.3.8) and (3.3.9), respectively, only the choice of the functions $\hat{g}, \tilde{g}$ and γ differs between the three models. We leave it to the reader to determine these functions for the models of Hart and Miller.

The model of Bruhns. To show that this constitutive model is of pre-monotone type, we set $N = 13$ and choose for the vector of internal variables $z = (\varepsilon_p, Y, \zeta) \in S^3 \times S^3 \times \mathbb{R} \cong \mathbb{R}^N$. The mapping B is chosen as in (3.3.8), whereas the linear mapping $M : S^3 \times S^3 \times \mathbb{R} \to S^3 \times S^3 \times \mathbb{R}$ is defined by

$$Mz = M(\hat{z}, \tilde{z}, \eta) = (D\hat{z}, \tilde{z}, \eta).$$

This mapping is symmetric and positive definite, and $L = M - B^T DB$ is positive semi-definite, since

$$(Mz) \cdot z = [M(\hat{z}, \tilde{z}, \eta)] \cdot (\hat{z}, \tilde{z}, \eta) = (D\hat{z}) \cdot \hat{z} + |\tilde{z}|^2 + \eta^2 > 0$$

for $z \neq 0$, and

$$(Lz) \cdot z = (0, \tilde{z}, \eta) \cdot (\hat{z}, \tilde{z}, \eta) = |\tilde{z}|^2 + \eta^2 \geq 0.$$

With the constants n, C_i and the function κ from (2.2.26) – (2.2.28) the function $g : S^3 \times S^3 \times \mathbb{R}^+ \to S^3 \times S^3 \times \mathbb{R}$ is defined by

$$g(z) = g(\hat{z}, \tilde{z}, \eta) = \left(\hat{g}(\hat{z}, \tilde{z}, \eta), \tilde{g}(\hat{z}, \tilde{z}, \eta), \gamma(\hat{z}, \tilde{z}, \eta) \right),$$

where

$$\hat{g}(\hat{z}, \tilde{z}, \eta) = C_1 \langle r \rangle \left(1 + \frac{r}{C_2}\right)^n \frac{P_0(\hat{z} + \tilde{z})}{|P_0(\hat{z} + \tilde{z})|}$$

$$\tilde{g}(\hat{z}, \check{z}, \eta) = \frac{3}{2}\left(C_3 - \frac{1}{2}\kappa'(-\eta)\right)\hat{g}(\hat{z}, \check{z}, \eta)$$

$$\gamma(\hat{z}, \check{z}, \eta) = \sqrt{\frac{3}{2}\,\kappa(-\eta)}\,C_1\langle r\rangle\left(1 + \frac{r}{C_2}\right)^n,$$

with $r = |P_0(\hat{z} + \check{z})| - \sqrt{\frac{3}{2}\kappa(-\eta)}$ and with $\langle r\rangle = \max(r, 0)$. It is immediately seen that with these definitions equations (2.2.24) – (2.2.28) can be written in the form

$$\rho u_{tt} = \operatorname{div}_x T$$
$$T = D(\varepsilon - Bz)$$
$$z_t = g(B^T D\varepsilon - Mz) = g(-\rho\nabla_z\psi(\varepsilon, z)),$$

where the free energy is given by

$$\rho\psi(\varepsilon, z) = \frac{1}{2}[D(\varepsilon - Bz)]\cdot(\varepsilon - Bz) + \frac{1}{2}(Lz)\cdot z$$

$$= \frac{1}{2}[D(\varepsilon - \varepsilon_p)]\cdot(\varepsilon - \varepsilon_p) + \frac{1}{2}|Y|^2 + \frac{1}{2}\zeta^2.$$

Here we set $z = (\varepsilon_p, Y, \zeta)$ to obtain the last equality sign. This proves that also the constitutive model of Bruhns is of pre-monotone type.

The model of Nouailhas and Freed. The proof, that this constitutive model is of pre-monotone type, differs little from the preceding ones. We choose $N = 12$. The vector of internal variables is $z = (\varepsilon_p, \varepsilon_n) \in \mathcal{S}^3 \times \mathcal{S}^3 \cong \mathbb{R}^N$. The mapping $B : \mathcal{S}^3 \times \mathcal{S}^3 \to \mathcal{S}^3$ is defined by

$$Bz = B(\hat{z}, \check{z}) = \hat{z} \in \mathcal{S}^3,$$

which implies that $B^T \hat{z} = (\hat{z}, 0)$, $B^T D\varepsilon = (D\varepsilon, 0)$, and $B^T DBz = B^T DB(\hat{z}, \check{z}) = (D\hat{z}, 0)$. For $M : \mathcal{S}^3 \times \mathcal{S}^3 \to \mathcal{S}^3 \times \mathcal{S}^3$ we choose

$$Mz = M(\hat{z}, \check{z}) = \left(D\hat{z} + \mathcal{N}(\hat{z} - \check{z}), -\mathcal{N}(\hat{z} - \check{z})\right),$$

where $\mathcal{N} : \mathcal{S}^3 \to \mathcal{S}^3$ is the linear, symmetric, positive definite mapping from (2.2.33). Since $\mathcal{N}$ is symmetric, also M is symmetric, and since $\mathcal{N}$ is positive definite, also M is positive definite and $L = M - B^T DB$ is positive semi-definite. For,

$$(Mz)\cdot z = [M(\hat{z}, \check{z})]\cdot(\hat{z}, \check{z}) = (D\hat{z})\cdot\hat{z} + [\mathcal{N}(\hat{z} - \check{z})]\cdot(\hat{z} - \check{z}) > 0$$

if $z \neq 0$ and

$$(Lz)\cdot z = [L(\hat{z}, \check{z})]\cdot(\hat{z}, \check{z}) = [\mathcal{N}(\hat{z} - \check{z})]\cdot(\hat{z} - \check{z}) \geq 0.$$

The mapping $g = \mathcal{S}^3 \times \mathcal{S}^3 \to \mathcal{S}^3 \times \mathcal{S}^3$ is defined by

$$g(z) = g(\hat{z}, \check{z}) = \left(\hat{g}(\hat{z}), \tilde{g}(\hat{z}, \check{z})\right)$$

with

$$\hat{g}(\hat{z}) = \left\langle\frac{|P_0\hat{z}|_R - k}{C}\right\rangle^n \frac{RP_0\hat{z}}{|P_0\hat{z}|_R}$$

$$\tilde{g}(\hat{z}, \check{z}) = \left\langle\frac{|P_0\hat{z}|_R - k}{C}\right\rangle^n \mathcal{N}^{-1}Q\hat{z}.$$

The equations (2.2.29) – (2.2.33) can now be written as

$$\rho u_{tt} = \operatorname{div}_x T$$
$$T = D(\varepsilon - Bz)$$
$$z_t = g(B^T D\varepsilon - Mz) = g(-\rho \nabla_z \psi(\varepsilon, z)),$$

with

$$\rho \psi(\varepsilon, z) = \frac{1}{2}[D(\varepsilon - Bz)] \cdot (\varepsilon - Bz) + \frac{1}{2}(Lz) \cdot z =$$

$$= \frac{1}{2}[D(\varepsilon - \varepsilon_p)] \cdot (\varepsilon - \varepsilon_p) + \frac{1}{2}[\mathcal{N}(\varepsilon_p - \varepsilon_n)] \cdot (\varepsilon_p - \varepsilon_n),$$

where we set $z = (\varepsilon_p, \varepsilon_n)$ to get the last equality sign.

The model of Méric, Poubanne and Cailletaud. To show that this crystallographic constitutive model is of pre-monotone type, we set $N = 3\nu$, where ν is the number of slip directions, and choose

$$z = (\varepsilon_p^1, \ldots, \varepsilon_p^\nu, \varepsilon_n^1, \ldots, \varepsilon_n^\nu, \zeta^1, \ldots, \zeta^\nu) \in \mathbf{R}^{3\nu} = \mathbf{R}^N$$

to be the vector of internal variables. With the matrices $m^s \in \mathcal{S}^3$, $s = 1, \ldots, \nu$, from (2.2.34), we define the mapping $B : \mathbf{R}^N \to \mathcal{S}^3$ by

$$Bz = B(\hat{z}^1, \ldots, \hat{z}^\nu, \tilde{z}^1, \ldots, \tilde{z}^\nu, \eta^1, \ldots, \eta^\nu) = \sum_{s=1}^{\nu} m^s \hat{z}^s. \qquad (3.3.11)$$

This yields for the transpose mapping $B^T : \mathcal{S}^3 \to \mathbf{R}^N$ that

$$B^T \sigma = (m^1 \cdot \sigma, \ldots, m^\nu \cdot \sigma, 0, \ldots, 0)$$
$$B^T D\varepsilon = (m^1 \cdot D\varepsilon, \ldots, m^\nu \cdot D\varepsilon, 0, \ldots, 0)$$
$$B^T DBz = B^T DB(\hat{z}^1, \ldots, \eta^\nu) = \qquad (3.3.12)$$

$$= \left(\sum_{s=1}^{\nu} m^1 \cdot Dm^s \hat{z}^s, \ldots, \sum_{s=1}^{\nu} m^\nu \cdot Dm^s \hat{z}^s, 0, \ldots, 0 \right).$$

For the linear mapping $M : \mathbf{R}^N \to \mathbf{R}^N$ we choose

$$Mz = M(\hat{z}^1, \ldots, \hat{z}^\nu, \tilde{z}^1, \ldots, \tilde{z}^\nu, \eta^1, \ldots, \eta^\nu) = \qquad (3.3.13)$$

$$= \left(\sum_{s=1}^{\nu} m^1 \cdot Dm^s \hat{z}^s + \mathcal{M}^1(\hat{z}^1 - \tilde{z}^1), \ldots, \sum_{s=1}^{\nu} m^\nu \cdot Dm^s \hat{z}^s + \mathcal{M}^\nu(\hat{z}^\nu - \tilde{z}^\nu), \right.$$

$$\left. -\mathcal{M}^1(\hat{z}^1 - \tilde{z}^1), \ldots, -\mathcal{M}^\nu(\hat{z}^\nu - \tilde{z}^\nu), \eta^1, \ldots, \eta^\nu \right),$$

where $\mathcal{M}^s > 0$ are the material constants from (2.2.42). It is immediately seen that M is symmetric.

We assume for the moment that the set $\{m^s\}_{s=1}^{\nu}$ is a linearly independent subset of $\mathcal{S}^3$, hence $\nu \le 6$. Under this assumption M is positive definite. To see this, note that (3.3.13) implies

$$(Mz) \cdot z = \left(\sum_{s=1}^{\nu} m^s \hat{z}^s \right) \cdot D\left(\sum_{s=1}^{\nu} m^s \hat{z}^s \right) + \sum_{s=1}^{\nu} \mathcal{M}^s\left(\hat{z}^s - \tilde{z}^s \right)^2 + \sum_{s=1}^{\nu} (\eta^s)^2. \qquad (3.3.14)$$

The right hand side of this equation is positive for $z \neq 0$. To see this note that $\mathcal{M}^s > 0$ for all s, that D is positive definite, and that $\sum_{s=1}^{\nu} m^s \hat{z}^s \neq 0$ for $(\hat{z}^1, \ldots, \hat{z}^\nu) \neq 0$. The last relation is a consequence of the assumed linear independence of the set $\{m^1, \ldots, m^\nu\}$.

Therefore M is positive definite. Moreover, $L = M - B^T D B$ is positive semi-definite, since (3.3.12) and (3.3.13) imply

$$(Lz) \cdot z = \sum_{s=1}^{\nu} \mathcal{M}^s \left(\hat{z}^s - \tilde{z}^s \right)^2 + \sum_{s=1}^{\nu} \left(\eta^s \right)^2 \geq 0 . \tag{3.3.15}$$

Finally, the mapping $g : \mathbb{R}^{2\nu} \times (0, C_4^1] \times \ldots \times (0, C_4^\nu] \to \mathbb{R}^N$ is defined by

$$g(z) = \left(\hat{g}^1(z), \ldots, \hat{g}^\nu(z), \tilde{g}^1(z), \ldots, \tilde{g}^\nu(z), \gamma^1(z), \ldots, \gamma^\nu(z) \right)$$

with

$$\hat{g}^s(z) = C_1^s \left\langle |\hat{z}^s| - k^s + \sum_{r=1}^{\nu} q_{sr} \eta^r \right\rangle^n \frac{\hat{z}^s}{|\hat{z}^s|}$$

$$\tilde{g}^s(z) = C_2^s |\hat{g}^s(z)| \, \tilde{z}^s$$

$$\gamma(z) = C_3^s |\hat{g}^s(z)| \, (C_4^s + \eta^s) .$$

One verifies immediately that with these definitions of B, M and g the equations (2.2.35) – (2.2.42) can be written as

$$\rho u_{tt} = \operatorname{div}_x T \tag{3.3.16}$$

$$T = D(\varepsilon - Bz) \tag{3.3.17}$$

$$z_t = g(B^T D\varepsilon - Mz) = g(-\rho \nabla_z \psi(\varepsilon, z)) , \tag{3.3.18}$$

where the free energy is

$$\rho \psi(\varepsilon, z) = \frac{1}{2} \left[D(\varepsilon - Bz) \right] \cdot (\varepsilon - Bz) + \frac{1}{2} (Lz) \cdot z =$$

$$= \frac{1}{2} \left[D \left(\varepsilon - \sum_{s=1}^{\nu} m^s \varepsilon_p^s \right) \right] \cdot \left(\varepsilon - \sum_{s=1}^{\nu} m^s \varepsilon_p^s \right) + \frac{1}{2} \sum_{s=1}^{\nu} \mathcal{M}^s \left(\varepsilon_p^s - \varepsilon_n^s \right)^2 + \frac{1}{2} \sum_{s=1}^{\nu} \left(\zeta^s \right)^2 .$$

To obtain the last equality, we set $z = (\varepsilon_p^1, \ldots, \varepsilon_p^\nu, \varepsilon_n^1, \ldots, \varepsilon_n^\nu, \zeta^1, \ldots, \zeta^\nu)$. This proves that the constitutive equations of Méric, Poubanne and Cailletaud are of pre-monotone type if the set $\{m^s\}_{s=1}^{\nu}$ is linearly independent.

Now we consider the general case and do not assume that this set is linearly independent. The equations (3.3.16) – (3.3.18) are equivalent to (2.2.35) – (2.2.42) also in this case, but from (3.3.14) we can now only conclude that M is positive semi-definite, which means that the equations (3.3.17), (3.3.18) do not necessarily satisfy the conditions of Definition 3.1.1. However, we can bring these equations in a form for which the conditions of Definition 3.1.1 are satisfied, and thus show that the constitutive model is also of pre-monotone type.

To this end we note first that the kernel of M is given by

$$\ker(M) = \{ z = (\hat{z}^1, \ldots, \eta^\nu) \in \mathbb{R}^N \mid \sum_{s=1}^{\nu} m^s \hat{z}^s = 0, \ \tilde{z}^s = \hat{z}^s, \ \eta^s = 0 \} . \tag{3.3.19}$$

This follows from the definition of M in (3.3.13), which shows that the linear space on the right hand side of (3.3.19) belongs to $\ker(M)$, and from (3.3.14), which shows that any element from $\ker(M)$ must belong to this linear space.

From this result and from the definition of B in (3.3.11) we see immediately that

$$\ker(M) \subseteq \ker(B).$$

For the ranges $R(M)$ of M and $R(B^T)$ of B^T we thus obtain

$$R(M) = [\ker(M^T)]^\perp = [\ker(M)]^\perp$$
$$R(B^T) = [\ker(B)]^\perp \subseteq [\ker(M)]^\perp = R(M).$$

Now let $P : \mathbb{R}^N \to \mathbb{R}^N$ be the orthogonal projection onto $R(M)$ along $\ker(M)$, hence $P(\mathbb{R}^N) = R(M)$, $P(\ker(M)) = \{0\}$, $(I - P)(\mathbb{R}^N) = \ker(M)$. Since $P^T = P$, we deduce that

$$M = M\big(P + (I - P)\big) = MP = PMP$$
$$B = B\big(P + (I - P)\big) = BP$$
$$B^T = (BP)^T = P^T B^T = PB^T.$$

We insert these equations into (3.3.17) and (3.3.18) to obtain the system

$$T = D(\varepsilon - BPz) \tag{3.3.20}$$
$$z_t = g(PB^T D\varepsilon - PMPz). \tag{3.3.21}$$

For the same initial data $z(0) = z^{(0)}$ and the same given function $t \mapsto \varepsilon(t)$, this system has the same solution $t \mapsto \big(z(t), T(t)\big)$ as the system (3.3.17), (3.3.18). Now, the right hand sides of both of the equations (3.3.20), (3.3.21) are independent of the component $(I - P)z$ of the vector z of internal variables. Therefore if we apply the mapping P to both sides of (3.3.21), we obtain a system for the component Pz of z alone:

$$T = D(\varepsilon - BPz) \tag{3.3.22}$$
$$(Pz)_t = Pg(PB^T D\varepsilon - PMPz). \tag{3.3.23}$$

For the initial data $Pz(0) = Pz^{(0)}$ and the function $t \mapsto \varepsilon(t)$, this system has the solution $t \mapsto \big(Pz(t), T(t)\big)$ with the same function $t \mapsto T(t)$ as in the solution of (3.3.17), (3.3.18). Therefore the systems (3.3.17), (3.3.18) and (3.3.22), (3.3.23) define the same mapping $(z^{(0)}, \varepsilon) \mapsto T$ and are thus equivalent systems of constitutive equations. We now identify the space $R(M)$ with $\mathbb{R}^\mu$, where $\mu = \dim R(M) < N$, and use the convention of writing w^* for the element from $\mathbb{R}^\mu$ identified with $w \in R(M)$. If we define the mappings $M^* : \mathbb{R}^\mu \to \mathbb{R}^\mu$, $B^* : \mathbb{R}^\mu \to \mathcal{S}^3$ and $g^* : \mathbb{R}^\mu \to \mathbb{R}^\mu$ by

$$M^* = PMP\big|_{R(M)} = M\big|_{R(M)}$$
$$B^* = BP\big|_{R(M)} = B\big|_{R(M)}$$
$$g^*(w^*) = Pg(w), \quad w^* \in D(g^*) \subseteq \mathbb{R}^\mu,$$

and write $z^*(x,t) = Pz(x,t)$ in (3.3.22), (3.3.23), and if we note that, with a slight abuse of notation,

$$\left(P|_{R(M)}\right)^T = P : \mathbb{R}^N \to R(M) \cong \mathbb{R}^\mu,$$

and hence

$$\left(B^*\right)^T = \left(BP|_{R(M)}\right)^T = \left(P|_{R(M)}\right)^T B^T = PB^T,$$

then we obtain the system of equations

$$\rho u_{tt} = \text{div}_x T \tag{3.3.24}$$

$$T = D(\varepsilon - B^* z^*) \tag{3.3.25}$$

$$z_t^* = g^*\left((B^*)^T D\varepsilon - M^* z^*\right) = g^*\left(-\rho \nabla_{z^*} \psi^*(\varepsilon, z^*)\right). \tag{3.3.26}$$

Here the free energy ψ^* is given by

$$\rho \psi^*(\varepsilon, z^*) = \frac{1}{2} \left[D(\varepsilon - B^* z^*)\right] \cdot (\varepsilon - B^* z^*) + \frac{1}{2} (L^* z^*) \cdot z^*$$

with $L^* = M^* - (B^*)^T D B^*$.

The mapping M is symmetric, whence also M^*. Moreover, (3.3.14) implies that M is positive semi-definite, which together with $\ker(M^*) = \ker(M|_{R(M)}) = \{0\}$ yields that M^* is positive definite. Finally, (3.3.15) shows that L is positive semi-definite, which implies that also $L^* = M^* - (B^*)^T D B^*$ is positive semi-definite, since

$$L^* = M^* - \left(B^*\right)^T D B^* = PMP|_{R(M)} - PB^T DBP|_{R(M)}$$

$$= P(M - B^T DB)P|_{R(M)} = PLP|_{R(M)}.$$

Consequently, the conditions of Definition 3.1.1 are satisfied by the constitutive equations (3.3.25), (3.3.26). Since the equations (3.3.24) – (3.3.26) are equivalent to the equations (2.2.35) – (2.2.42), it follows that the constitutive model of Méric, Poubanne and Cailletaud is of pre-monotone type also in the case where the set $\left\{m^s\right\}_{s=1}^\nu$ is not linearly independent.

Alternatively, describing the preceding calculation more precisely, one can say that if $\left\{m^s\right\}_{s=1}^\nu$ is not linearly independent, then the model contains too many interior variables. After elimination of the unnecessary variables an equivalent system of constitutive equations results, which is of pre-monotone type.

3.4 A Criterion for Monotone Type

We conclude this chapter by stating a criterion which allows us to determine whether constitutive equations are of monotone type, and which will be used in Chapter 8 to study engineering models. For simplicity we only consider rate-dependent constitutive equations in the following lemma.

For a function $f : D(f) \subseteq S^3 \times \mathbb{R}^N \to \mathbb{R}^N$ let

$$D_0(f) = \{z \in \mathbb{R}^N \mid (0, z) \in D(f)\}.$$

Lemma 3.4.1 *Let the constitutive equations*

$$T = D(\varepsilon(\nabla_x u) - Bz) \tag{3.4.1}$$

$$z_t = f(\varepsilon(\nabla_x u), z) \tag{3.4.2}$$

be given.
(I) *If a symmetric, positive definite $N \times N$-matrix M exists satisfying the conditions*

(i) $L = M - B^T DB$ *is positive semi-definite*

(ii) $\nabla_\varepsilon f(\varepsilon, z) + \nabla_z f(\varepsilon, z) M^{-1} \overline{M} = 0$ \hfill (3.4.3)

for all $(\varepsilon, z) \in D(f)$, where $\overline{M} = B^T D : \mathcal{S}^3 \to \mathbb{R}^N$,

(iii) $-\nabla_z f(0, z) M^{-1}$ *is a positive semi-definite $N \times N$-matrix for all $z \in D_0(f)$,*

(iv) $f(0, z) \cdot Mz \leq 0$ *for all $z \in D_0(f)$,*

and if in addition the domain of f satisfies the two conditions

(v) $D_0(f)$ *is convex,*

(vi) $(\varepsilon_0, z_0) + V_M \subseteq D(f)$

for all $(\varepsilon_0, z_0) \in D(f)$, where V_M denotes the subspace

$$V_M = \{(\varepsilon, z) \in \mathcal{S}^3 \times \mathbb{R}^N \mid z = M^{-1} \overline{M} \varepsilon\}$$

of $\mathcal{S}^3 \times \mathbb{R}^N$, then the constitutive equations (3.4.1), (3.4.2) are of monotone type. Moreover, f can be represented in the form

$$f(\varepsilon, z) = g(-\rho \nabla_z \psi(\varepsilon, z)), \tag{3.4.4}$$

where the free energy ψ and the monotone vector field g are defined by

$$\rho \psi(\varepsilon, z) = \frac{1}{2} [D(\varepsilon - Bz)] \cdot (\varepsilon - Bz) + \frac{1}{2} (Lz) \cdot z \tag{3.4.5}$$

$$g(z) = f(0, -M^{-1} z), \quad z \in D(g) = -M(D_0(f)). \tag{3.4.6}$$

This free energy satisfies $-\rho \nabla_z \psi(\varepsilon, z) = \overline{M}\varepsilon - Mz$.

(II) *Conversely, if the constitutive equations (3.4.1), (3.4.2) are of monotone type, which means that f can be represented in the form (3.4.4) with a suitable monotone vector field g and a suitable free energy ψ, then there exists a symmetric, positive definite $N \times N$-matrix M (namely the matrix defined by $Mz = \rho \nabla_z \psi(0, z)$) satisfying the conditions (i) - (iv).*

Proof. We remark first that $f(\varepsilon, z)$ is constant on every connected component of the submanifold $[(\varepsilon_0, z_0) + V_M] \cap D(f)$ for all (ε_0, z_0) if and only if (3.4.3) holds in $D(f)$. Since $\bar{\varepsilon} \mapsto (\bar{\varepsilon} + \varepsilon_0, M^{-1} \overline{M} \bar{\varepsilon} + z_0)$ is a bijective map of $\mathcal{S}^3$ onto $(\varepsilon_0, z_0) + V_M$, the function $f(\varepsilon, z)$ is constant on connected components of $[(\varepsilon_0, z_0) + V_M] \cap D(f)$

if and only if the map $\bar{\varepsilon} \mapsto f(\bar{\varepsilon} + \varepsilon_0, M^{-1}\overline{M}\bar{\varepsilon} + z_0)$, whose domain is a subset of $\mathcal{S}^3$, is constant on connected components, and thus if and only if

$$0 = \nabla_{\bar{\varepsilon}} f(\bar{\varepsilon} + \varepsilon_0, M^{-1}\overline{M}\bar{\varepsilon} + z_0)$$

$$= \nabla_\varepsilon f(\bar{\varepsilon} + \varepsilon_0, M^{-1}\overline{M}\bar{\varepsilon} + z_0) + \nabla_z f(\bar{\varepsilon} + \varepsilon_0, M^{-1}\overline{M}\bar{\varepsilon} + z_0) M^{-1}\overline{M}$$

for all $\bar{\varepsilon} \in \mathcal{S}^3$ with $(\bar{\varepsilon} + \varepsilon_0, M^{-1}\overline{M}\bar{\varepsilon} + z_0) \in D(f)$. Equivalent to this requirement is that (3.4.3) holds for all $(\varepsilon, z) \in D(f)$, since (ε_0, z_0) can be chosen arbitrarily.

To prove (I), first note that because of (i) the quadratic form ψ defined in (3.4.5) is positive semi-definite. Consequently, to show that (f, B) is of monotone type, it suffices to prove that (3.4.4) holds with ψ and g defined in (3.4.5), (3.4.6), that this g is monotone, and that it satisfies $g(z) \cdot z \geq 0$. To prove (3.4.4) we use that for $(\varepsilon, z) \in D(f)$ condition (ii) implies that the function f is constant on every connected component of the set $[(\varepsilon, z) + V_M] \cap D(f)$. Now, condition (vi) implies that this set is equal to $(\varepsilon, z) + V_M$, hence f is constant on $(\varepsilon, z) + V_M$. Since $(-\varepsilon, -M^{-1}\overline{M}\varepsilon) \in V_M$, we consequently obtain

$$f(\varepsilon, z) = f(0, z - M^{-1}\overline{M}\varepsilon) = f(0, -M^{-1}(\overline{M}\varepsilon - Mz))$$

$$= g(\overline{M}\varepsilon - Mz) = g(-\rho\nabla_z\psi(\varepsilon, z)),$$

where the third equality follows from the definition (3.4.6) of g, and the last equality from the definition of ψ in (3.4.5). Therefore (3.4.4) is fulfilled.

We next show that g is monotone. From (3.4.6) and (iii) we obtain that the matrix

$$\nabla g(z) = -\nabla_z f(0, -M^{-1}z) M^{-1}$$

is positive semi-definite. By (v), $D_0(f)$ and consequently also $D(g) = -M(D_0(f))$ is convex. For $z, \bar{z} \in D(g)$ and $0 \leq t \leq 1$ we thus have $t(\bar{z} - z) + z \in D(g)$, whence

$$[g(\bar{z}) - g(z)] \cdot (\bar{z} - z) = \left[\int_0^1 \frac{d}{dt} g(t(\bar{z} - z) + z) dt \right] \cdot (\bar{z} - z)$$

$$= \int_0^1 \left[\nabla g(t(\bar{z} - z) + z)(\bar{z} - z) \right] \cdot (\bar{z} - z) dt \geq 0,$$

where in the last step we used that the integrand is non-negative, due to the fact that ∇g is positive semi-definite. This inequality implies that g is monotone, and since (3.4.6) and (iv) yield with $\bar{z} = -M^{-1}z$ that

$$g(z) \cdot z = f(0, -M^{-1}z) \cdot z = -f(0, \bar{z}) \cdot M\bar{z} \geq 0 \qquad (3.4.7)$$

for all $z \in D(g) = -M(D_0(f))$, it follows that (f, B) is of monotone type.

To prove (II), assume that the constitutive equations (3.4.1), (3.4.2) are of monotone type. Then f satisfies (3.4.4) with a suitable monotone vector field g and a suitable free energy ψ, which by Definition 3.1.1 has the property that the linear mapping M defined by $Mz = \rho\nabla_z\psi(0, z)$ is symmetric, positive definite and satisfies (i). To prove that this M satisfies (3.4.3), note that for $(\varepsilon, z) \in (\varepsilon_0, z_0) + V_M$ we have $z - z_0 = M^{-1}\overline{M}(\varepsilon - \varepsilon_0)$, hence

$$-\rho\nabla_z\psi(\varepsilon, z) = \overline{M}\varepsilon - Mz = \overline{M}(\varepsilon - \varepsilon_0) - M(z - z_0) + \overline{M}\varepsilon_0 - Mz_0$$

$$= \overline{M}(\varepsilon - \varepsilon_0) - MM^{-1}\overline{M}(\varepsilon - \varepsilon_0) + \overline{M}\varepsilon_0 - Mz_0 = \overline{M}\varepsilon_0 - Mz_0.$$

This implies that $\rho\nabla_z\psi(\varepsilon,z)$, whence $g(-\rho\nabla_z\psi(\varepsilon,z))$, and consequently, because of (3.4.4), $f(\varepsilon,z)$ are constant on $(\varepsilon_0,z_0)+V_M$. The remarks at the beginning of this proof thus show that (3.4.3) holds. To see that (iii) holds, note that (3.4.4) yields

$$-\nabla_z f(0,z)M^{-1} = \nabla g(-Mz).$$

For the proof of (iii) it thus suffices to show that the matrix $\nabla g(z)$ is positive semi-definite for all $z \in D(g)$. This follows immediately from the monotonicity of g, since for $\bar{z} \in \mathbb{R}^N$ and $t \in \mathbb{R}\backslash\{0\}$

$$0 \le \frac{1}{t^2}[g(z+t\bar{z})-g(z)]\cdot t\bar{z} = \frac{1}{t^2}[\nabla g(z)t\bar{z}+o(1)t]\cdot t\bar{z}$$

$$= [\nabla g(z)\bar{z}]\cdot\bar{z}+o(1), \quad t\to 0,$$

whence $[\nabla g(z)\bar{z}]\cdot\bar{z} \ge 0$, and thus $\nabla g(z)$ is positive semi-definite. The proof of condition (iv) is obtained by reading (3.4.7) backwards.

The proof of Lemma 3.4.1 is complete.

Chapter 4

Existence of Solutions for Constitutive Equations of Monotone Type

4.1 Formulation of the Problem as a First Order Evolution Equation

In this chapter we use the theory of evolution equations for monotone operators to prove that the initial-boundary value problem

$$\rho u_{tt} = \operatorname{div}_x T$$
$$T = D(\varepsilon(\nabla_x u) - Bz)$$
$$z_t \in f(\varepsilon(\nabla_x u), z)$$
$$u(x,t)\big|_{x \in \partial\Omega} = 0 \quad \text{or} \quad T(x,t)n(x)\big|_{x \in \partial\Omega} = 0$$
$$u(x,0) = u^{(0)}(x),\, u_t(x,0) = u^{(1)}(x),\, z(x,0) = z^{(0)}(x)$$

has a unique solution if the constitutive equations are of monotone type and satisfy two additional conditions. Namely, we require that

$$f(\varepsilon, z) = g(-\rho \nabla_z \psi(\varepsilon, z)),$$

for an operator

$$g : \mathbb{R}^N \to \mathcal{P}(\mathbb{R}^N),$$

which is not only monotone, but maximal monotone, and a quadratic form

$$\rho\psi(\varepsilon, z) = \frac{1}{2}[D(\varepsilon - Bz)] \cdot (\varepsilon - Bz) + \frac{1}{2}(Lz) \cdot z,$$

which is not only positive semi-definite, but positive definite. More precisely, we show that the following equivalent problem obtained from this initial-boundary value problem by setting $v = u_t$, $e = \varepsilon(\nabla_x u)$ has a unique solution:

$$v_t(x,t) = \operatorname{div} \frac{1}{\rho} D\big(e(x,t) - Bz(x,t)\big) \tag{4.1.1}$$

$$e_t(x,t) = \varepsilon(\nabla_x v(x,t)) \tag{4.1.2}$$

$$z_t(x,t) \in g\big(-\rho \nabla_z \psi(e(x,t), z(x,t))\big) \tag{4.1.3}$$

$$v(x,t)\big|_{x \in \partial\Omega} = 0 \tag{4.1.4}$$

or

$$[D(e(x,t) - Bz(x,t)]n(x)\big|_{x \in \partial\Omega} = 0 \tag{4.1.5}$$

$$v(x,0) = v^{(0)}(x), \; e(x,0) = e^{(0)}(x), \; z(x,0) = z^{(0)}(x), \; x \in \Omega. \tag{4.1.6}$$

Here $\varepsilon(\nabla_x v(x,t))$ is defined in (2.1.4). The solution is a mapping

$$(v,e,z) : \Omega \times [0,\infty) \to \mathbb{R}^3 \times \mathcal{S}^3 \times \mathbb{R}^N. \tag{4.1.7}$$

v is the rate of deformation and e is the strain tensor.

Our first goal is to write the problem (4.1.1) – (4.1.6) in the form

$$w_t(t) \in C(w(t)), \quad w(0) = w^{(0)} = (v^{(0)}, e^{(0)}, z^{(0)}) \tag{4.1.8}$$

of an initial value problem for an evolution equation with a maximal monotone operator $-C$, and with

$$w(t) = (v(\cdot,t), e(\cdot,t), z(\cdot,t)) : \Omega \to \mathbb{R}^3 \times \mathcal{S}^3 \times \mathbb{R}^N.$$

To this end we introduce some notations. As usual, we define the domain of the set valued map g by

$$D(g) = \{z \in \mathbb{R}^N \mid g(z) \neq \emptyset\}.$$

By $H_m(\Omega, \mathbb{R}^n)$ we denote the Sobolev space of all functions from $L^2(\Omega, \mathbb{R}^n)$, which have weak derivatives in $L^2(\Omega, \mathbb{R}^n)$ up to order m. The scalar product and norm on this space are denoted by

$$(u,v)_m = (u,v)_{m,\Omega} = \sum_{|\alpha| \leq m} \int_\Omega D^\alpha u(x) \cdot D^\alpha v(x) dx,$$

and

$$\|u\|_m = \|u\|_{m,\Omega} = (u,u)^{1/2}_{m,\Omega}.$$

We also set $(u,v) = (u,v)_0$ and $\|u\| = \|u\|_0$. $\overset{\circ}{H}_m(\Omega)$ denotes the closure of the space of test functions $C_0^\infty(\Omega)$ in $H_m(\Omega)$. If for $u \in L^2(\Omega, \mathbb{R}^3)$ there is a function $v \in L^2(\Omega, \mathbb{R})$ such that

$$(v,w) = -(u, \nabla w)$$

holds for all $w \in \overset{\circ}{H}_1(\Omega, \mathbb{R})$, then we say that the weak divergence of u exists and is given by v. Since the weak divergence v is uniquely defined, we denote it by $\operatorname{div} u$.

We now define two operators C_D and C_N corresponding to the Dirichlet boundary condition (4.1.4) and the Neumann boundary condition (4.1.5). To this end we set for brevity

$$W = \mathbb{R}^3 \times \mathcal{S}^3 \times \mathbb{R}^N$$

and first define an operator

$$C_0 : L^2(\Omega, W) \to \mathcal{P}(L^2(\Omega, W))$$

by $C_0(v, e, z) = \emptyset$ if $v \in L^2(\Omega, \mathbb{R}^3) \backslash H_1(\Omega, \mathbb{R}^3)$, and

$$C_0(v, e, z) =$$

$$\{w = (w_1, w_2, w_3) \in L^2(\Omega, W) \mid w_1 = \text{div} \frac{1}{\rho} D(e - Bz), \tag{4.1.9}$$

$$w_2 = \varepsilon(\nabla v), w_3(x) \in g\big(-\rho\nabla_z\psi(e(x), z(x))\big) \text{ a.e. in } \Omega\}$$

if $(v, e, z) \in H_1(\Omega, \mathbb{R}^3) \times L^2(\Omega, \mathcal{S}^3 \times \mathbb{R}^N)$. Note that this definition implies that the domain

$$D(C_0) = \{(v, e, z) \mid C_0(v, e, z) \neq \emptyset\}$$

consists of all

$$(v, e, z) \in H_1(\Omega, \mathbb{R}^3) \times L^2(\Omega, \mathcal{S}^3 \times \mathbb{R}^N),$$

for which the weak divergence div $\frac{1}{\rho} D(e - Bz)$ exists, and for which there is $w_3 \in L^2(\Omega, \mathbb{R}^N)$ with

$$w_3(x) \in g\big(-\rho\nabla_z\psi(e(x), z(x))\big)$$

for almost all $x \in \Omega$.

The operators

$$C_D, C_N : L^2(\Omega, W) \to \mathcal{P}(L^2(\Omega, W))$$

are now obtained from C_0 by restriction of the domain. For the Dirichlet problem we choose as domain

$$D(C_D) = \{(v, e, z) \in D(C_0) \mid v \in \overset{\circ}{H}_1(\Omega, \mathbb{R}^3)\}, \tag{4.1.10}$$

whereas for the Neumann problem the domain $D(C_N)$ consists of all $(v, e, z) \in D(C_0)$ for which the weak divergence satisfies

$$(\text{div} \frac{1}{\rho} D(e - Bz), u) = -(\frac{1}{\rho} D(e - Bz), \nabla u) \tag{4.1.11}$$

for all $u \in H_1(\Omega, \mathbb{R}^3)$. If C denotes C_D or C_N, we define

$$C(v, e, z) = \begin{cases} C_0(v, e, z), & \text{if } (v, e, z) \in D(C) \\ \emptyset, & \text{if } (v, e, z) \in L^2(\Omega, W) \backslash D(C). \end{cases} \tag{4.1.12}$$

Also in the remainder of this chapter we use the convention of writing C if a statement holds for C_D and C_N, and the symbol C can therefore always be replaced by C_D or C_N.

A weak formulation of the initial-boundary value problem (4.1.1) – (4.1.6) is now given by (4.1.8). The boundary conditions (4.1.4) and (4.1.5) are contained in the definitions of C_D and C_N, since if $(v, e, z) \in D(C_D)$, then $v \in \overset{\circ}{H}_1(\Omega)$, and if $(v, e, z) \in D(C_N)$, then (4.1.11) is satisfied, which is a weak form of the Neumann boundary condition (4.1.5).

4.2   Maximality of the Evolution Operator

To apply the theory of evolution equations for monotone operators we next show that $-C$ is a maximal monotone operator.

Theorem 4.2.1 *Let the quadratic form*

$$\rho\psi(e,z) = \frac{1}{2}[D(e-Bz)] \cdot (e-Bz) + \frac{1}{2}(Lz) \cdot z \qquad (4.2.1)$$

be positive definite on $\mathcal{S}^3 \times \mathbb{R}^N$ and let $g : \mathbb{R}^N \to \mathcal{P}(\mathbb{R}^N)$ be maximal monotone with $0 \in g(0)$. Let $\Omega \subseteq \mathbb{R}^3$ be an open set if $C = C_D$, and let Ω be an open, bounded set with Lipschitz boundary if $C = C_N$. Then the operator $-C : L^2(\Omega, W) \to \mathcal{P}(L^2(\Omega, W))$ is maximal monotone, if the space $L^2(\Omega, W)$ is equiped with the scalar product

$$\langle (v,e,z), (\overline{v}, \overline{e}, \overline{z}) \rangle = \int_\Omega \rho v \cdot \overline{v} + [D(e-Bz)] \cdot (\overline{e} - B\overline{z}) + (Lz) \cdot \overline{z}\, dx\,. \qquad (4.2.2)$$

Remarks In the following proof we need Korn's first inequality if $C = C_D$ and Korn's second inequality if $C = C_N$. Korn's second inequality holds in the domain Ω if this domain is bounded and has Lipschitz boundary. This is the reason for the assumption on Ω in this theorem.

A monotone operator $A : \mathcal{H} \to \mathcal{P}(\mathcal{H})$ on a Hilbert space $\mathcal{H}$ is called maximal monotone if it does not have a proper monotone extension, cf. [23, p. 22]. As an example of a maximal monotone operator $g : \mathbb{R}^N \to \mathcal{P}(\mathbb{R}^N)$, consider a monotone vector field $\overline{g} : \mathbb{R}^N \to \mathbb{R}^N$ which is continuous and defined on all of $\mathbb{R}^N$. Then the single-valued operator $g(z) = \{\overline{g}(z)\}$ generated by $\overline{g}$ is maximal monotone, cf. [23, p. 33].

We note that if $\rho\psi(e,z)$ is positive definite on $\mathcal{S}^3 \times \mathbb{R}^N$, then the bilinear form defined by (4.2.2) is a scalar product, because

$$\langle (v,e,z), (v,e,z) \rangle = \int_\Omega \rho|v|^2 + 2\rho\psi(e,z)dx > 0$$

for $(v,e,z) \neq 0$. Since for every $z \in \mathbb{R}^N$ there exists $e \in \mathcal{S}^3$ with $e - Bz = 0$, and since for this e

$$\frac{1}{2}(Lz) \cdot z = \frac{1}{2}[D(e-Bz)] \cdot (e-Bz) + \frac{1}{2}(Lz) \cdot z = \rho\psi(e,z)\,,$$

it follows that $\rho\psi$ is positive definite if and only if the symmetric $N \times N$–matrix L in (4.2.1) is positive definite.

Proof of Theorem 4.2.1   I.) In the first step of the proof we show that $-C$ is monotone. To this end it must be verified that

$$\langle w - \overline{w}, (v,e,z) - (\overline{v}, \overline{e}, \overline{z}) \rangle \leq 0$$

for all $(v,e,z), (\overline{v}, \overline{e}, \overline{z}) \in D(C)$ and $w \in C(v,e,z), \overline{w} \in C(\overline{v}, \overline{e}, \overline{z})$. We set $w^* = w - \overline{w}$ and $(v^*, e^*, z^*) = (v,e,z) - (\overline{v}, \overline{e}, \overline{z})$. Then $w^* = (w_1^*, w_2^*, w_3^*) \in \mathbb{R}^3 \times \mathcal{S}^3 \times \mathbb{R}^N$, and the definition of C in (4.1.9) – (4.1.12) implies

$$w_1^* = w_1 - \overline{w}_1 = \operatorname{div} \frac{1}{\rho} D(e - Bz) - \operatorname{div} \frac{1}{\rho} D(\overline{e} - B\overline{z}) = \operatorname{div} \frac{1}{\rho} D(e^* - Bz^*),$$

$$w_2^* = w_2 - \overline{w}_2 = \varepsilon(\nabla_x v) - \varepsilon(\nabla_x \overline{v}) = \varepsilon(\nabla_x v^*),$$

which yields

$$\langle w - \overline{w}, (v, e, z) - (\overline{v}, \overline{e}, \overline{z}) \rangle$$

$$= \int_{\Omega} \rho w_1^* \cdot v^* + [D(w_2^* - Bw_3^*)] \cdot (e^* - Bz^*) + (Lw_3^*) \cdot z^* dx$$

$$= \int_{\Omega} \rho (\mathrm{div} \frac{1}{\rho} D(e^* - Bz^*)) \cdot v^* + (\varepsilon(\nabla v^*) - Bw_3^*) \cdot [D(e^* - Bz^*)]$$

$$+ (Lw_3^*) \cdot z^* \, dx \qquad (4.2.3)$$

$$= \int_{\Omega} -[D(e^* - Bz^*)] \cdot \nabla v^* + [D(e^* - Bz^*)] \cdot \varepsilon(\nabla v^*)$$

$$- \left(Bw_3^* \cdot [D(e^* - Bz^*)] - w_3^* \cdot Lz^* \right) dx \,,$$

where in the case $C = C_D$ the partial integration is allowed because of the definition of the weak divergence and since $v^* \in \overset{\circ}{H}_1(\Omega, \mathbb{R}^3)$, and in the case $C = C_N$ because of (4.1.11) and since $v \in H_1(\Omega, \mathbb{R}^3)$. We also used that D and L are symmetric linear mappings on $\mathcal{S}^3$ and on $\mathbb{R}^N$, respectively. We finally use that $T = D(e^* - Bz^*) \in \mathcal{S}^3$, hence

$$T \cdot \varepsilon(\nabla v^*) = T \cdot \frac{1}{2}(\nabla v^* + (\nabla v^*)^T)$$

$$= \frac{1}{2}[T \cdot (\nabla v^*) + T^T \cdot (\nabla v^*)^T] = T \cdot (\nabla v^*) = D(e^* - Bz^*) \cdot \nabla v^*.$$

Insertion of this equation into (4.2.3) and application of (3.1.4) results in

$$\langle w - \overline{w}, (v, e, z) - (\overline{v}, \overline{e}, \overline{z}) \rangle$$

$$= - \int_{\Omega} Bw_3^* \cdot [D(e^* - Bz^*)] - w_3^* \cdot Lz^* dx$$

$$= - \int_{\Omega} w_3^* \cdot [B^T D(e^* - Bz^*) - Lz^*] dx$$

$$= - \int_{\Omega} w_3^* \cdot [-\rho \nabla_z \psi(e^*, z^*)] dx$$

$$= - \int_{\Omega} (w_3 - \overline{w}_3) \cdot [-\rho \nabla_z \psi(e, z) + \rho \nabla_z \psi(\overline{e}, \overline{z})] dz \leq 0 \,,$$

where we also used by definition of C in (4.1.9), that,

$$w_3(x) \in g\left(-\rho \nabla_z \psi(e(x), z(x))\right), \quad \overline{w}_3(x) \in g\left(-\rho \nabla_z \psi(\overline{e}(x), \overline{z}(x))\right)$$

for almost all $x \in \Omega$. The inequality thus follows from the monotonicity of g. This concludes the proof of the monotonicity of $-C$.

II.) In the second step of the proof we show that $-C$ is maximal montone. As usual, for a monotone operator $A : \mathcal{H} \to \mathcal{P}(\mathcal{H})$ on a Hilbert space $\mathcal{H}$ and for $\lambda > 0$ we define the operator

$$(\lambda I + A) : \mathcal{H} \to \mathcal{P}(\mathcal{H})$$

by

$$(\lambda I + A)w = \lambda w + Aw$$

for $w \in D(\lambda I + A) = D(A)$. The range $R(\lambda I + A)$ is given by

$$R(\lambda I + A) = \bigcup_{w \in D(A)} (\lambda I + A)w,$$

and the inverse

$$(\lambda I + A)^{-1} : \mathcal{H} \to \mathcal{P}(\mathcal{H})$$

by

$$\overline{w} \in (\lambda I + A)^{-1}w \iff w \in (\lambda I + A)\overline{w}.$$

The operator $-C$ is maximal monotone if and only if for a $\lambda > 0$ we have

$$R(\lambda I - C) = L^2(\Omega, W),$$

cf. [23, p. 23]. Therefore $-C$ is maximal monotone if and only if for a $\lambda > 0$ the relation

$$r \in (\lambda I - C)w \tag{4.2.4}$$

has a solution w for every $r \in L^2(\Omega, W)$. We remark that $(\lambda I - C)^{-1}$ is single-valued, since $-C$ is monotone, which implies that the solution w of (4.2.4) must be unique.

Consequently, to complete the proof of Theorem 4.2.1 it suffices to show that (4.2.4) has a solution $w = (v, e, z) \in D(C)$ for every $r = (r_1, r_2, r_3) \in L^2(\Omega, \mathbb{R}^3 \times \mathcal{S}^3 \times \mathbb{R}^N)$. To prove the existence of this solution, we use the definition of C in (4.1.9) – (4.1.12) to write (4.2.4) in the explict form

$$-\operatorname{div} \frac{1}{\rho} D(e(x) - Bz(x)) + \lambda v(x) = r_1(x) \tag{4.2.5}$$

$$-\varepsilon(\nabla v(x)) + \lambda e(x) = r_2(x) \tag{4.2.6}$$

$$-g(-\rho \nabla_z \psi(e(x), z(x))) + \lambda z(x) \ni r_3(x), \tag{4.2.7}$$

where the solution (v, e, z) must belong to the space $\overset{\circ}{H}_1(\Omega, \mathbb{R}^3) \times L^2(\Omega, \mathcal{S}^3 \times \mathbb{R}^N)$ for $C = C_\mathcal{D}$, whereas for $C = C_\mathcal{N}$ the solution (v, e, z) must belong to the space $H_1(\Omega, \mathbb{R}^3) \times L^2(\Omega, \mathcal{S}^3 \times \mathbb{R}^N)$ and the function $D(e - Bz)$ must satisfy (4.1.11).

IIa.) To construct the solution of (4.2.5) – (4.2.7), we first study the relation (4.2.7) and show that if $\lambda > 0$ then for every $r_3 \in L^2(\Omega, \mathbb{R}^N)$ and every $e \in L^2(\Omega, \mathcal{S}^3)$ there is a unique $z \in L^2(\Omega, \mathbb{R}^N)$ satisfying (4.2.7).

To this end we consider the relation

$$-g(-\rho \nabla_z \psi(e, z)) + \lambda z \ni r_3 \tag{4.2.8}$$

on $\mathbb{R}^N$ and prove that for every $r_3 \in \mathbb{R}^N$ and $e \in \mathcal{S}^3$ there is a unique $z \in \mathbb{R}^N$, which satisfies (4.2.8) and depends continuously on r_3 and e. Using (3.1.4) we rewrite (4.2.8) as

$$r_3 \in \lambda z - g(B^T De - Mz), \tag{4.2.9}$$

with the symmetric, positive definite linear mapping $M = L + B^T DB : \mathbb{R}^N \to \mathbb{R}^N$. If we insert h defined by

$$h = B^T De - Mz,$$

into (4.2.9), we obtain

$$-r_3 + \lambda M^{-1} B^T De \in g(h) + \lambda M^{-1} h. \tag{4.2.10}$$

h is a solution of this equation if and only if z given by

$$z = M^{-1} B^T De - M^{-1} h$$

is a solution of (4.2.9) and therefore of (4.2.8). Thus, to prove that (4.2.8) has a solution, it suffices to show that (4.2.10) has a solution.

To prove that (4.2.10) is solvable, note that there exists $\mu > 0$ such that $(\lambda M^{-1} - \mu I)$ is positive definite and therefore monotone, since λM^{-1} is positive definite. As linear mapping on a finite dimensional space, $(\lambda M^{-1} - \mu I)$ is Lipschitz continuous. The sum of the maximal monotone operator g and the monotone, Lipschitz continuous mapping $(\lambda M^{-1} - \mu I)$ is therefore a maximal monotone operator

$$g + (\lambda M^{-1} - \mu I) : \mathbb{R}^N \to \mathbb{R}^N,$$

cf. [23, p. 34]. Since μ is positive, it thus follows that

$$g + \lambda M^{-1} = [g + (\lambda M^{-1} - \mu I)] + \mu I$$

is surjective, cf. [23, p. 23], and consequently (4.2.10) has a solution h and therefore (4.2.8) a solution z for all given r_3, e.

To see that the solution is unique and depends continuously on r_3 and e, let $z, \bar{z}$ be solutions of (4.2.8) for r_3, e and $\bar{r}_3, \bar{e}$, respectively. We then have

$$-r_3 + \lambda z \in g(-\rho \nabla_z \psi(e, z)), \quad -\bar{r}_3 + \lambda \bar{z} \in g(-\rho \nabla_z \psi(\bar{e}, \bar{z})).$$

The monotonicity of g thus implies

$$0 \leq [-\rho \nabla_z \psi(e, z) + \rho \nabla_z \psi(\bar{e}, \bar{z})] \cdot [(-r_3 + \lambda z) - (-\bar{r}_3 + \lambda \bar{z})]$$

$$= [(B^T De - Mz) - (B^T D\bar{e} - M\bar{z})] \cdot [(-r_3 + \lambda z) - (-\bar{r}_3 + \lambda \bar{z})] \tag{4.2.11}$$

$$= -[\lambda M(z - \bar{z})] \cdot (z - \bar{z})$$

$$+ [M(z - \bar{z})] \cdot (r_3 - \bar{r}_3) + [\lambda B^T D(e - \bar{e})] \cdot (z - \bar{z}) - [B^T D(e - \bar{e})] \cdot (r_3 - \bar{r}_3).$$

Since M is positive definite, there exists $\nu > 0$ with $(M\zeta) \cdot \zeta \geq \nu |\zeta|^2$ for all $\zeta \in \mathbb{R}^N$, hence

$$\lambda \nu |z - \bar{z}|^2 \leq [\lambda M(z - \bar{z})] \cdot (z - \bar{z})$$

$$\leq (\|M\| \, |r_3 - \bar{r}_3| + \lambda \|B^T D\| \, |e - \bar{e}|) \, |z - \bar{z}| + \|B^T D\| \, |e - \bar{e}| \, |r_3 - \bar{r}_3|$$

$$\leq \frac{1}{2}\lambda \nu \, |z - \bar{z}|^2 + \frac{1}{2\lambda \nu}(\|M\| \, |r_3 - \bar{r}_3| + \lambda \|B^T D\| \, |e - \bar{e}|)^2$$

$$+ \|B^T D\| \, |e - \bar{e}| \, |r_3 - \bar{r}_3|,$$

$$\leq \frac{\lambda \nu}{2}|z - \bar{z}|^2 + \frac{1}{2}\Big[\frac{1}{\lambda \nu}(\|M\| + \lambda \|B^T D\|)^2 + \|B^T D\|\Big](|e - \bar{e}| + |r_3 - \bar{r}_3|)^2,$$

whence

$$|z - \bar{z}| \leq K(|e - \bar{e}| + |r_3 - \bar{r}_3|) \tag{4.2.12}$$

with

$$K = \left(\frac{1}{\lambda\nu}\Big[\frac{1}{\lambda\nu}(\|M\| + \lambda\|B^T D\|)^2 + \|B^T D\|\Big]\right)^{1/2}.$$

From this equation we obtain $z = \bar{z}$ if $(\bar{e}, \bar{r}_3) = (e, r_3)$, which shows that the solution z of (4.2.8) is unique. By setting

$$G(e, r_3) = z,$$

we can therefore define a function $G : \mathcal{S}^3 \times \mathbb{R}^N \to \mathbb{R}^N$, which assigns the solution z of (4.2.8) to (e, r_3). Using (4.2.12) again, we conclude that G is uniformly Lipschitz continuous. Moreover, since by assumption $0 \in g(0)$, it is obvious that $(\bar{e}, \bar{z}, \bar{r}_3) = 0$ satisfies (4.2.8). Thus, setting $(\bar{e}, \bar{z}, \bar{r}_3) = 0$ in (4.2.12) yields the estimate

$$|G(e, r_3)| \leq K(|e| + |r_3|) \tag{4.2.13}$$

for G.

We need a second estimate for G. To derive it, we consider the case $r_3 = \bar{r}_3$ and set $z = G(e, r_3)$, $\bar{z} = G(\bar{e}, r_3)$, $e^* = e - \bar{e}$, $z^* = z - \bar{z}$. From (4.2.11) we then deduce

$$(Mz^*) \cdot z^* \leq (B^T De^*) \cdot z^* = (De^*) \cdot Bz^*,$$

which together with $M = L + B^T DB$ yields

$$(Mz^*) \cdot z^* = (Lz^* + B^T DBz^*) \cdot z^*$$
$$= (Lz^*) \cdot z^* + (DBz^*) \cdot Bz^* \leq (De^*) \cdot Bz^*. \tag{4.2.14}$$

Since $\rho\psi$ and therefore L are positive definite, there exists $\nu_1 > 0$ with

$$(L\zeta) \cdot \zeta \geq \nu_1 |\zeta|^2$$

for all $\zeta \in \mathbb{R}^N$, from which we conclude that

$$\frac{\nu_1}{1 + \|B^T DB\|}(DBz^*) \cdot Bz^*$$
$$= \frac{\nu_1}{1 + \|B^T DB\|}(B^T DBz^*) \cdot z^* \leq \nu_1 |z^*|^2 \leq (Lz^*) \cdot z^*. \tag{4.2.15}$$

We use now that D is symmetric and positive definite on $\mathcal{S}^3$, which implies that $(D\zeta) \cdot \eta$ defines a scalar product on $\mathcal{S}^3$. Therefore we can apply the Cauchy-Schwarz inequality, which together with (4.2.14) and (4.2.15) yields for $\gamma = \nu_1/(1 + \|B^T DB\|)$ that

$$(1 + \gamma)(DBz^*) \cdot Bz^* \leq (DBz^*) \cdot Bz^* + (Lz^*) \cdot z^*$$
$$\leq (De^*) \cdot Bz^* \leq [(De^*) \cdot e^*]^{1/2}[(DBz^*) \cdot Bz^*]^{1/2},$$

hence

$$\big[[D(Bz - B\bar{z})] \cdot (Bz - B\bar{z})\big]^{1/2} \leq \frac{1}{1 + \gamma}\big[[D(e - \bar{e})] \cdot (e - \bar{e})\big]^{1/2}, \tag{4.2.16}$$

where $z = G(e, r_3), \overline{z} = G(\overline{e}, r_3)$. This is the second estimate for G.

To conclude this part of the proof, let $e \in L^2(\Omega, \mathcal{S}^3)$, $r_3 \in L^2(\Omega, \mathbb{R}^N)$ be given and define the function $z : \Omega \to \mathbb{R}^N$ by

$$z(x) = G(e(x), r_3(x)).$$

Since e and r_3 are measurable and since G is Lipschitz continuous, it follows that z is measurable, and from (4.2.13) we obtain

$$|z(x)| \le K(|e(x)| + |r_3(x)|),$$

which implies $z \in L^2(\Omega, \mathbb{R}^N)$ with norm

$$\|z\| \le K(\|e\| + \|r_3\|).$$

By definition of G we also have

$$-r_3(x) \in g\big(-\rho \nabla_z \psi(e(x), z(x))\big) - \lambda z(x)$$

for almost all $x \in \Omega$. Therefore z is a solution of (4.2.7). Finally, if

$$|e|_\Omega = \left(\int_\Omega (De(x)) \cdot e(x) dx \right)^{1/2}$$

denotes the norm belonging to the scalar product $\int_\Omega (De(x)) \cdot \overline{e}(x) dx$ on $L^2(\Omega, \mathcal{S}^3)$, we obtain from (4.2.16)

$$|BG(e, r_3) - BG(\overline{e}, r_3)|_\Omega \le \frac{1}{1+\gamma} |e - \overline{e}|_\Omega \tag{4.2.17}$$

with $\gamma > 0$, where $G(e, r_3)$ denotes the function $x \mapsto G(e(x), r_3(x))$.

IIb.) We next show that for all $\lambda > 0$, $r_1 \in L^2(\Omega, \mathbb{R}^3)$, $z \in L^2(\Omega, \mathbb{R}^N)$ and $r_2 \in L^2(\Omega, \mathcal{S}^3)$ there exists a unique function pair

$$(v, e) \in \begin{cases} \overset{\circ}{H}_1(\Omega, \mathbb{R}^3) \times L^2(\Omega, \mathcal{S}^3), & \text{if } C = C_\mathcal{D} \\ H_1(\Omega, \mathbb{R}^3) \times L^2(\Omega, \mathcal{S}^3), & \text{if } C = C_\mathcal{N}, \end{cases}$$

which solves (4.2.5) and (4.2.6), and which for $C = C_\mathcal{N}$ satisfies the weak Neumann boundary condition (4.1.11). We also show that if $r_1 = r_2 = 0$, then this solution satisfies

$$|e|_\Omega \le |Bz|_\Omega. \tag{4.2.18}$$

To prove these statements, note first that from the definition of the weak divergence it follows that $e, Bz \in L^2(\Omega, \mathcal{S}^3)$ and $v, r_1 \in L^2(\Omega, \mathbb{R}^3)$ satisfy (4.2.5) if and only if

$$-\int_\Omega \frac{1}{\rho} D(e - Bz) \cdot \nabla u \, dx = \int_\Omega (\lambda v - r_1) \cdot u \, dx \tag{4.2.19}$$

holds for all $u \in \overset{\circ}{H}_1(\Omega, \mathbb{R}^3)$. In addition, the boundary condition (4.1.11) is satisfied if and only if (4.2.19) holds for all $u \in H_1(\Omega, \mathbb{R}^3)$. The equation (4.2.6) can be solved for e. The resulting equation is

$$e = \frac{1}{\lambda}(\varepsilon(\nabla v) + r_2). \tag{4.2.20}$$

If we insert the right hand side of this equation into (4.2.19) and use that the symmetry of $D(e - Bz)$ implies

$$D(e - Bz) \cdot \nabla u = D(e - Bz) \cdot \varepsilon(\nabla u),$$

we obtain

$$\int_\Omega \frac{1}{\lambda \rho} [D\varepsilon(\nabla v)] \cdot \varepsilon(\nabla u) + \lambda v \cdot u \, dx \qquad (4.2.21)$$

$$= \int_\Omega \frac{1}{\rho} \Big[D\big(Bz - \frac{1}{\lambda} r_2 \big) \Big] \cdot \nabla u + r_1 \cdot u \, dx.$$

Now we have reduced the solution of (4.2.5), (4.2.6) to the solution of (4.2.21). Namely, it is immediately seen that

$$(v, e) \in \begin{cases} \overset{\circ}{H}_1(\Omega, \mathbb{R}^3) \times L^2(\Omega, \mathcal{S}^3), & \text{if } C = C_\mathcal{D} \\ H_1(\Omega, \mathbb{R}^3) \times L^2(\Omega, \mathcal{S}^3), & \text{if } C = C_\mathcal{N} \end{cases}$$

satisfies (4.2.20) and (4.2.21) for all

$$u \in \begin{cases} \overset{\circ}{H}_1(\Omega, \mathbb{R}^3), & \text{if } C = C_\mathcal{D} \\ H_1(\Omega, \mathbb{R}), & \text{if } C = C_\mathcal{N} \end{cases}$$

if and only if (v, e) satisfies (4.2.19) and (4.2.20). In turn this means that (4.2.5), (4.2.6) and the boundary conditions are fulfilled if and only if v is a solution of (4.2.21) and for this v the function e is defined by (4.2.20).

However, the existence and uniqueness of the solution of (4.2.21) follows as usual from the Theorem of Lax-Milgram, since for every

$$(r_1, r_2, Bz) \in L^2(\Omega, \mathbb{R}^3 \times (\mathcal{S}^3)^2)$$

the right hand side of (4.2.21) defines a continuous linear functional

$$u \mapsto \int_\Omega \frac{1}{\rho} [D(Bz - \frac{1}{\lambda} r_2)] \cdot \nabla u + r_1 \cdot u \, dx,$$

and the bilinear form on the left hand side is bounded and coercive on $\overset{\circ}{H}_1(\Omega, \mathbb{R}^3)$ or $H_1(\Omega, \mathbb{R}^3)$, respectively. The coercivity follows from Korn's inequality

$$\|u\|_{1,\Omega}^2 \leq K_1 (\|u\|_{0,\Omega}^2 + \|\varepsilon(\nabla u)\|_{0,\Omega}^2).$$

If $\Omega \subseteq \mathbb{R}^3$ is open, then this inequality holds for all $u \in \overset{\circ}{H}_1(\Omega, \mathbb{R}^3)$, and if Ω is a bounded domain with Lipschitz boundary, then this inequality holds for all $u \in H_1(\Omega, \mathbb{R}^3)$, cf.[172, p. 13 ff]. Therefore, since in the case $C = C_\mathcal{N}$ we assumed that Ω is a bounded Lipschitz domain, in both the Dirichlet and Neumann case we obtain from Korn's inequality and from the assumption that D is positive definite that $\nu_2 > 0$ exists with

$$\frac{\nu_2}{K_1} \|u\|_{1,\Omega}^2 \leq \int_\Omega \nu_2 (|\varepsilon(\nabla u)|^2 + |u|^2) dx$$

$$\leq \int_\Omega \frac{1}{\lambda \rho} [D\varepsilon(\nabla u)] \cdot \varepsilon(\nabla u) + \lambda |u|^2 dx$$

for all $u \in \overset{\circ}{H}_1(\Omega, \mathbb{R}^3)$ or $u \in H_1(\Omega, \mathbb{R}^3)$, respectively. This showes that the bilinear form is coercive.

Thus we proved that (4.2.5), (4.2.6) have a unique solution (v, e) satisfying the corresponding boundary conditions.

If $r_1 = r_2 = 0$, then we obtain for this solution from (4.2.20), (4.2.21) that

$$\lambda \int_\Omega \frac{1}{\rho}(De) \cdot e + |v|^2 dx = \int_\Omega \frac{1}{\rho}(DBz) \cdot \nabla v \, dx$$

$$= \int_\Omega \frac{1}{\rho}(DBz) \cdot \varepsilon(\nabla v) dx = \lambda \int_\Omega \frac{1}{\rho}(DBz) \cdot e \, dx,$$

where we set $u = v$ and again used the symmetry of DBz to get the second equality. Multiplication of this inequality by ρ/λ and application of the Cauchy-Schwarz inequality yields

$$\int_\Omega (De) \cdot e \, dx \leq \left(\int_\Omega (DBz) \cdot Bz \, dx \right)^{1/2} \left(\int_\Omega (De) \cdot e \, dx \right)^{1/2} = |Bz|_\Omega |e|_\Omega,$$

which proves (4.2.18).

IIc.) We define the linear operator

$$\mathcal{R} : L^2(\Omega, \mathbb{R}^N \times \mathbb{R}^3 \times \mathcal{S}^3) \to L^2(\Omega, \mathcal{S}^3)$$

by setting

$$\mathcal{R}(z, r_1, r_2) = e,$$

where e is the second component of the function pair (v, e), which solves (4.2.5), (4.2.6) for z, r_1, r_2, and which satisfies the Dirichlet or Neumann boundary condition. From (4.2.18) we obtain

$$|\mathcal{R}(z, 0, 0)|_\Omega \leq |Bz|_\Omega. \tag{4.2.22}$$

For given $r = (r_1, r_2, r_3) \in L^2(\Omega, \mathbb{R}^3 \times \mathcal{S}^3 \times \mathbb{R}^N)$ let the mapping $\mathcal{P} : L^2(\Omega, \mathcal{S}^3) \to L^2(\Omega, \mathcal{S}^3)$ be defined by

$$\mathcal{P}(\bar{e}) = \mathcal{R}(G(\bar{e}, r_3), r_1, r_2).$$

It follows from (4.2.17) and (4.2.22) that $\mathcal{P}$ is a contraction mapping, since the linearity of $\mathcal{R}$ implies for $e, \bar{e} \in L^2(\Omega, \mathcal{S}^3)$ that

$$|\mathcal{P}(e) - \mathcal{P}(\bar{e})|_\Omega = |\mathcal{R}(G(e, r_3), r_1, r_2) - \mathcal{R}(G(\bar{e}, r_3), r_1, r_2)|_\Omega$$

$$= |\mathcal{R}(G(e, r_3) - G(\bar{e}, r_3), 0, 0)|_\Omega \leq |BG(e, r_3) - BG(\bar{e}, r_3)|_\Omega$$

$$\leq \frac{1}{1+\gamma}|e - \bar{e}|_\Omega$$

with $1/(1+\gamma) < 1$. Thus, the contraction $\mathcal{P}$ has a unique fixed point $e^* \in L^2(\Omega, \mathcal{S}^3)$. Now define z by

$$z = G(e^*, r^3),$$

and let (v, e) be the function pair, which satisfies (4.2.5), (4.2.6) for z, r_1, r_2, and which satisfies the weak Dirichlet or Neumann boundary condition. Then

$$e = \mathcal{R}(z, r_1, r_2) = \mathcal{R}(G(e^*, r_3), r_1, r_2) = \mathcal{P}(e^*) = e^*,$$

which implies that v, e^*, z, r_1, r_2 satisfy (4.2.5), (4.2.6) and the Dirichlet or Neumann boundary condition, and e^*, z, r_3 satisfy (4.2.7). Thus $w = (v, e^*, z)$ is a solution of (4.2.5) – (4.2.7) and of the boundary conditions, and therefore of (4.2.4) for $r = (r_1, r_2, r_3)$. It follows that $\lambda I - C$ is surjective for every $\lambda > 0$, which proves that $-C$ is maximal monotone. The proof of Theorem 4.2.1 is complete.

4.3   Existence for the Dynamic Problem

Since $-C$ is maximal monotone we can now apply the theory of evolution equations to monotone operators to prove existence and uniqueness of solutions for the initial-boundary value problem (4.1.8), the weak formulation of the the initial-boundary value problem (4.1.1) – (4.1.6), and immediately obtain the following result, cf. [23, p. 54]:

Theorem 4.3.1 Let $C = C_D$ or $C = C_N$ be the operator on $L^2(\Omega, \mathbb{R}^3 \times \mathcal{S}^3 \times \mathbb{R}^N)$ defined in (4.1.9) – (4.1.12), and let the hypotheses of Theorem 4.2.1 be satisfied. Then to every $w^{(0)} = (v^{(0)}, e^{(0)}, z^{(0)}) \in D(C)$ there exists a unique function $w : [0, \infty) \to L^2(\Omega, \mathbb{R}^3 \times \mathcal{S}^3 \times \mathbb{R}^N)$, $w(t) = (v(\cdot, t), e(\cdot, t), z(\cdot, t))$, such that

(i) $w(t) \in D(C)$ for all $t \geq 0$

(ii) w is Lipschitz continuous on $[0, \infty)$

(iii) $w_t(t) \in C(w(t))$, almost everywhere in $[0, \infty)$

(iv) $w(0) = w^{(0)}$.

Moreover, w has the following properties:

(v) w has everywhere a derivative from the right.

(vi)   *If w and $\overline{w}$ are two solutions satisfying the conditions* (i) – (iii), *then*

$$\|w(t) - \overline{w}(t)\|_\psi \leq \|w(0) - \overline{w}(0)\|_\psi$$

for all $t \geq 0$.

Here

$$\|(v, e, z)\|_\psi = \left(\int_\Omega \rho |v(x)|^2 + 2\rho\psi(e(x), z(x)) dx \right)^{1/2}$$

is the norm of $L^2(\Omega, \mathbb{R}^3 \times \mathcal{S}^3 \times \mathbb{R}^N)$ defined by the scalar product (4.2.2).

More detailed information than we stated can be derived from the general theory of evolution equations of monotone operators, and in particular, the existence of weak solutions can be shown, which allows to solve the initial-boundary value problem (4.1.8) for a larger class of initial data. For these results we refer the reader to [23].

Chapter 5

Transformation of Interior Variables

The existence theorem of the preceding chapter applies to initial-boundary value problems with constitutive equations of monotone type. It is possible to enlarge the scope of this theorem to a class of constitutive equations which are not of monotone type by transformation of interior variables. For a certain class of constitutive equations it is possible to choose this transformation such that the new system of constitutive equations satisfied by the transformed interior variables is of monotone type, and thus allows us to apply the existence theorem from the preceding chapter. In this and in the following chapter we study this transformation for rate dependent constitutive equations, and derive conditions which characterize the class of constitutive equations transformable to monotone type and to the other types introduced in Definition 3.1.1. The transformation of rate independent constitutive equations is investigated in Chapter 7.

5.1 Transformation Fields

We study the transformation of interior variables for the system

$$\rho u_{tt} = \text{div}_x T \tag{5.1.1}$$

$$T = D(\varepsilon(\nabla_x u) - Bz) \tag{5.1.2}$$

$$z_t = f(\varepsilon(\nabla_x u), z), \tag{5.1.3}$$

where $\varepsilon(\nabla_x u)$ is defined in (2.1.4), where

$$B : \mathbb{R}^N \to \mathcal{S}^3$$

is a linear mapping, and where

$$f : D(f) \subseteq \mathcal{S}^3 \times \mathbb{R}^N \to \mathbb{R}^N.$$

We suppose that

$$(u, z) : D((u, z)) \subseteq \Omega \times [0, \infty) \to \mathbb{R}^3 \times \mathbb{R}^N.$$

Let $H : D(H) \subseteq \mathbb{R}^N \to \mathbb{R}^N$ be a continuously differentiable vector field and let

$$h : D(h) \subseteq \Omega \times [0, \infty) \to \mathbb{R}^N, \quad h(x,t) = H(z(x,t))$$

be the transformed interior variables. We call H the transformation field and, for simplicity, we suppose in this work that $D(H) = \mathbb{R}^N$ and that H is bijective on $\mathbb{R}^N$ with continuously differentiable inverse H^{-1}. By

$$H'(z) = \Big(\frac{\partial}{\partial z_j} H_i(z) \Big)_{i,j=1,\dots,N}$$

we denote the $N \times N$–matrix of first partial derivatives of H. From our assumptions for H it follows that $H'(z)$ is non singular.

Lemma 5.1.1 *If (u, z) is a solution of (5.1.1) – (5.1.3), then the function (u, h) satisfies the transformed system*

$$\rho u_{tt} = \operatorname{div}_x T \tag{5.1.4}$$
$$T = D(\varepsilon(\nabla_x u) - BH^{-1}(h)) \tag{5.1.5}$$
$$h_t = f_H(\varepsilon(\nabla_x u), h) \tag{5.1.6}$$

with the transformed function $f_H : D(f_H) \subseteq \mathcal{S}^3 \times \mathbb{R}^N \to \mathbb{R}^N$ defined by

$$f_H(\varepsilon, h) = H'(H^{-1}(h)) f(\varepsilon, H^{-1}(h)). \tag{5.1.7}$$

Proof. We have

$$\frac{\partial}{\partial t} h(x,t) = \frac{\partial}{\partial t} H(z(x,t)) = H'(z(x,t)) \frac{\partial}{\partial t} z(x,t)$$
$$= H'(z(x,t)) f(\varepsilon(x,t), z(x,t))$$
$$= H'\Big(H^{-1}(h(x,t)) \Big) f\big(\varepsilon(x,t), H^{-1}(h(x,t)) \big) = f_H(\varepsilon(x,t), h(x,t)).$$

This computation shows that (5.1.6) is satisfied; (5.1.4) and (5.1.5) immediately follow from (5.1.1) and (5.1.2).

Let (f, B) be the pair consisting of the function f and the linear mapping B from the right hand side of the constitutive equations (5.1.2), (5.1.3). Lemma 5.1.1 shows that the transformation field H associates to this pair the pair (f_H, BH^{-1}) from the transformed constitutive equations (5.1.5), (5.1.6). Therefore H induces the transformation $\mathcal{T}_H : (f, B) \to (f_H, BH^{-1})$ of pairs. However, in this work we only consider pairs, whose second argument is a linear mapping. As consequence, the map $BH^{-1} : \mathbb{R}^N \to \mathcal{S}^3$ must be linear, which restricts the domain of definition of the transformation $\mathcal{T}_H$ to a subset of the set of all pairs of functions $f : D(f) \subseteq \mathcal{S}^3 \times \mathbb{R}^N \to \mathbb{R}^N$ and linear mappings $B : \mathbb{R}^N \to \mathcal{S}^3$. This subset consists of those pairs (f, B), for which BH^{-1} is linear.

Our goal is to determine H and therefore the transformation $\mathcal{T}_H$ such that the constitutive equations (5.1.5), (5.1.6) are of monotone type. Then we can take advantage of the properties of monotone operators in the investigation of the system (5.1.4) - (5.1.6), and thus also in the investigation of the original system (5.1.1) - (5.1.3), since a solution of (5.1.4) - (5.1.6) can be transformed back to a solution of the original system by the vector field H^{-1}. If, for example, the transformed system is of monotone type and satisfies the stronger hypotheses needed in Theorems 4.2.1 and 4.3.1, namely that the vector field g is maximal monotone and the free energy is positive definite, then the initial-boundary value problem to the transformed system (5.1.4) – (5.1.6) has a unique solution (u, h), and this solution exists globally in time. A globally existing solution (u, z) of the initial-boundary value problem of the original system (5.1.1) – (5.1.3) is then obtained by transformation of (u, h) with the vector field H^{-1}.

Thus, the problem arises to characterize the class of constitutive equations that can be transformed to monotone type, and to find the transformation field which performs this transformation for a given constitutive equation. We study this transformation problem in the remainder of this chapter, and start with the following

Definition 5.1.2 *Let* $\mathcal{TM}^*$, $\mathcal{TG}$, $\mathcal{TM}$ *and* $\mathcal{T(MG)}$, *respectively, denote the set of pairs* (f, B) *of functions* $f : D(f) \subseteq \mathcal{S}^3 \times \mathbb{R}^N \to \mathbb{R}^N$ *and linear mappings* $B : \mathbb{R}^N \to \mathcal{S}^3$ *with the property, that the constitutive equations*

$$T = D(\varepsilon - Bz)$$
$$z_t = f(\varepsilon, z)$$

can be transformed to pre-monotone type, to gradient type, to monotone type, or to monotone-gradient type, respectively.

Obviously the relations

$$\mathcal{T(MG)} \subseteq \mathcal{T(G \cap M)} \subseteq \mathcal{TG} \cap \mathcal{TM}, \quad \mathcal{TG}, \mathcal{TM} \subseteq \mathcal{TM}^*$$

hold, since, as an immediate consequence of Definition 3.1.1, the classes $\mathcal{M}^*$, $\mathcal{G}$, $\mathcal{M}$ and $\mathcal{MG}$ satisfy $\mathcal{MG} \subseteq (\mathcal{G} \cap \mathcal{M})$ and $\mathcal{G}$, $\mathcal{M} \subseteq \mathcal{M}^*$.

In the following corollary we formulate conditions, which characterize the classes $\mathcal{TM}^*$, $\mathcal{TG}$, $\mathcal{TM}$ and $\mathcal{T(MG)}$, and which therefore can be used to decide whether a pair (f, B) belongs to one of these classes. These conditions are not very explicit, though, and the main goal of the investigations in the remainder of this chapter is to derive more explicit conditions.

If the pair (f, B) is transformed by the vector field H, then according to Definition 3.1.1, the transformed pair (f_H, BH^{-1}) is of pre-monotone type if and only if

$$BH^{-1} : \mathbb{R}^N \to \mathcal{S}^3$$

is linear and f_H satisfies

$$f_H(\varepsilon, h) = g(-\rho \nabla_h \psi(\varepsilon, h)) \tag{5.1.8}$$

for all $(\varepsilon, h) \in D(f_H)$, for a suitable function g and for a suitable quadratic, positive semi-definite free energy ψ. If we insert (5.1.7) into (5.1.8) and transfer the conditions for g and ψ given in Definition 3.1.1 to the system (5.1.4) – (5.1.6), we immediately obtain the following result:

Corollary 5.1.3 *The pair* (f, B) *belongs to the class* $\mathcal{T}\mathcal{M}^*$ *if and only if there exist a vector field* $H : \mathbb{R}^N \to \mathbb{R}^N$, *a continuous vector field* $g : D(g) \subseteq \mathbb{R}^N \to \mathbb{R}^N$, *and a symmetric, positive definite* $N \times N$*-matrix* M *such that*

$$BH^{-1} : \mathbb{R}^N \to \mathcal{S}^3 \qquad (5.1.9)$$

is a linear mapping, such that

$$L = M - (BH^{-1})^T D (BH^{-1}) \qquad (5.1.10)$$

is positive semi-definite, and such that

$$H'(H^{-1}(h)) f(\varepsilon, H^{-1}(h)) = g(-\rho \nabla_h \psi(\varepsilon, h)) \qquad (5.1.11)$$

holds for all $(\varepsilon, h) \in D(f_H) = \{(\bar{\varepsilon}, H(z)) \mid (\bar{\varepsilon}, z) \in D(f)\}$, *where*

$$\rho \psi(\varepsilon, h) = \frac{1}{2} [D(\varepsilon - BH^{-1}(h))] \cdot (\varepsilon - BH^{-1}(h)) + \frac{1}{2} (Lh) \cdot h. \qquad (5.1.12)$$

This vector field H *transforms the system* (5.1.1) - (5.1.3) *to pre-monotone type. The transformed system is given by*

$$\rho u_{tt} = \operatorname{div}_x T$$

$$T = D(\varepsilon(\nabla_x u) - BH^{-1}(h))$$

$$h_t = g(-\rho \nabla_h \psi(\varepsilon(\nabla_x u), h)).$$

The pair (f, B) *belongs to the class* $\mathcal{T}\mathcal{G}$ *if and only if* H, g *and* M *satisfy these conditions and* g *is equal to the gradient* $\nabla \chi$ *of a proper function* $\chi : D(\chi) \subseteq \mathbb{R}^N \to (-\infty, \infty]$. *The pair* (f, B) *belongs to* $\mathcal{T}\mathcal{M}$ *if and only these conditions for* H, g, M *are fulfilled and* g *is a monotone vector field satisfying* $g(z) \cdot z \geq 0$ *for all* $z \in D(g)$. *Finally, the pair* (f, B) *belongs to* $\mathcal{T}(\mathcal{M}\mathcal{G})$ *if and only if these conditions for* H, g, M *are fulfilled and* g *satisfies* $g(z) \cdot z \geq 0$ *for all* $z \in D(g)$ *and is equal to the gradient of a proper convex function.*

5.2 Properties of the Transformation: Restrictions Imposed by the ε–Independence of the Transformation Field and Invariance under Linear Transformations

If the pair (f, B) belongs to one of the classes $\mathcal{T}\mathcal{M}^*, \mathcal{T}\mathcal{G}, \mathcal{T}\mathcal{M}$, or $\mathcal{T}(\mathcal{M}\mathcal{G})$, then vector fields H and g exist satisfying (5.1.11). If g is given, then (5.1.11) is a system of N first order partial differential equations for the N components of H. This does not mean however, that H can be determined from this system to any given function g, since this system contains a parameter ε, and (5.1.11) must hold for all possible values of ε, whereas the solutions H and g must be independent of ε. This requirement imposes an implicit restriction on the vector field H. If H were allowed to depend on ε, then the transformed equation (5.1.6) would contain the term $\varepsilon(\nabla_x u_t)$, which we want to avoid. This problem is discussed more thorougly in Section 9.1.

We next study this implicit restriction and in the following lemma formulate it as an explicit condition for H. To this end we use that (5.1.12) yields

$$\rho \nabla_h \psi(\varepsilon, h) = -(BH^{-1})^T D(\varepsilon - BH^{-1}(h)) + LH = -\overline{M}\varepsilon + Mh \qquad (5.2.1)$$

with the symmetric, positive definite $N \times N$–matrix M from Corollary 5.1.3 and with the linear mapping

$$\overline{M} = (BH^{-1})^T D : \mathcal{S}^3 \to \mathbb{R}^N. \qquad (5.2.2)$$

Since M is invertible, equation (5.2.1) implies that for every $c \in \mathbb{R}^N$ the set

$$\{(\varepsilon, h) \in \mathcal{S}^3 \times \mathbb{R}^N \mid \rho \nabla_h \psi(\varepsilon, h) = c\} = \{(\varepsilon, h) \mid -\overline{M}\varepsilon + Mh = c\}$$

is a six-dimensional submanifold of $\mathcal{S}^3 \times \mathbb{R}^N$, and consequently also the inverse image

$$\begin{aligned} V(c, M, H) &= \{(\varepsilon, z) \in \mathcal{S}^3 \times \mathbb{R}^N \mid \rho \nabla_h \psi(\varepsilon, H(z)) = c\} \\ &= \{(\varepsilon, z) \in \mathcal{S}^3 \times \mathbb{R}^N \mid -\overline{M}\varepsilon + MH(z) = c\} \end{aligned}$$

of this submanifold under the mapping $(\varepsilon, z) \mapsto (\varepsilon, H(z))$. $V(c, M, H)$ contains a unique element of the form $(0, z)$, namely the element with $z = H^{-1}(M^{-1}c)$. Therefore the intersection of $V(c, M, H)$ with the set $\{0\} \times D_0(f)$ is empty or consists of the point $(0, H^{-1}(M^{-1}c))$, where

$$D_0(f) = \{z \in \mathbb{R}^N \mid (0, z) \in D(f)\}.$$

Lemma 5.2.1 (I) *If equation (5.1.11) holds, then the following two conditions are satisfied:*

(i)
$$H'(z) f(0, z) = g\big(-\rho \nabla_h \psi(0, H(z))\big) \qquad (5.2.3)$$

for all $z \in D_0(f)$

(ii)
$$H'(z) f(\varepsilon, z) = \text{const}$$

on every set

$$V(c, M, H) \cap D(f) = \{(\varepsilon, z) \in \mathcal{S}^3 \times \mathbb{R}^N \mid \rho \nabla_h \psi(\varepsilon, H(z)) = c\} \cap D(f).$$

Conversely, if (i) and (ii) hold and if for every c the submanifold $V(c, M, H)$ intersects $\{0\} \times D_0(f) \subseteq \mathcal{S}^3 \times \mathbb{R}^N$ whenever it intersects $D(f)$, then (5.1.11) holds.

(II) *If condition (ii) holds, then the system*

$$\begin{aligned} H'(z)[\nabla_\varepsilon f(\varepsilon, z) + \nabla_z f(\varepsilon, z) H'(z)^{-1} M^{-1} \overline{M}] \\ + H''(z) : f(\varepsilon, z) H'(z)^{-1} M^{-1} \overline{M} = 0 \end{aligned} \qquad (5.2.4)$$

of partial differential equations is satisfied for all $(\varepsilon, z) \in D(f)$, where $H''(z)$ denotes the tensor

$$H''(z) = (H_{ij}^k(z))_{i,j,k=1,\dots,N},$$

with

$$H_{ij}^k(z) = \frac{\partial^2}{\partial z_k \partial z_j} H_i(z),$$

and where

$$H''(z) : f(0, z) = \Big(\sum_{k=1}^N H_{ij}^k(z) f_k(z) \Big)_{i,j=1,\dots,N}.$$

Conversely, if (5.2.4) is satisfied and if every one of the sets $V(c, M, H) \cap D(f)$ is connected, then condition (ii) holds.

Since we assumed that H is bijective, (5.2.3) is of course equivalent to

$$H'(H^{-1}(h)) f(0, H^{-1}(h)) = g(-\rho \nabla_h \psi(0, h)).$$

This lemma thus shows that if condition (ii) can be guaranteed, then it suffices to satisfy equation (5.1.11) for $\varepsilon = 0$. The dependence of this equation on the parameter ε is thus removed, but the solution H is now explicitly restricted by the additional condition (ii) or by (5.2.4). For every (ε, z) the left hand side of (5.2.4) is a linear mapping on $\mathcal{S}^3$. Therefore (5.2.4) can be considered to be a system of six partial differential equations.

Proof. (I) Suppose that conditions (i) and (ii) hold, that $V(c, M, H)$ intersects $\{0\} \times D_0(f)$ whenever it intersects $D(f)$, and that $(\varepsilon, z) \in D(f)$ is given. For

$$c = -\overline{M}\varepsilon + M H(z)$$

we have $(\varepsilon, z) \in V(c, M, H)$. Consequently there exists $\overline{z}$ with $(0, \overline{z}) \in V(c, M, H) \cap D(f)$. We thus obtain

$$\begin{aligned} H'(z) f(\varepsilon, z) &= H'(\overline{z}) f(0, \overline{z}) = g\Big(-\rho \nabla_h \psi(0, H(\overline{z}))\Big) \\ &= g\Big(-\rho \nabla_h \psi(\varepsilon, H(z))\Big), \end{aligned}$$

which proves (5.1.11). If on the other hand (5.1.11) holds, then (5.2.3) is obtained by setting $\varepsilon = 0$. To prove that condition (ii) is satisfied, let $(\varepsilon, z), (\overline{\varepsilon}, \overline{z}) \in V(c, M, H) \cap D(f)$. Then (5.1.11) yields

$$\begin{aligned} H'(z) f(\varepsilon, z) &= g\Big(-\rho \nabla_h \psi(\varepsilon, H(z))\Big) \\ &= g\Big(-\rho \nabla_h \psi(\overline{\varepsilon}, H(\overline{z}))\Big) = H'(\overline{z}) f(\overline{\varepsilon}, \overline{z}), \end{aligned}$$

hence $H'(z) f(\varepsilon, z)$ is constant on $V(c, M, H) \cap D(f)$. This proves (ii).

(II) To verify (5.2.4), assume that (ii) holds. Then

$$V(c, M, H) = \Big\{ \big(\varepsilon, H^{-1}(M^{-1}(c + \overline{M}\varepsilon))\big) \mid \varepsilon \in \mathcal{S}^3 \Big\}$$

implies that the mapping

$$\varepsilon \mapsto H'\Big(H^{-1}(M^{-1}(c + \overline{M}\varepsilon))\Big) f\Big(\varepsilon, H^{-1}(M^{-1}(c + \overline{M}\varepsilon))\Big) \qquad (5.2.5)$$

is constant. Therefore the left hand side of (5.2.4) vanishes, since it is equal to the derivative of this function. On the other hand, if (5.2.4) holds, then the function defined in (5.2.5) is constant on connected components of its domain. This means that $H'(z) f(\varepsilon, z)$ is constant on connected components of the set $V(c, M, H) \cap D(f)$, and consequently on this whole set if it is connected. Therefore condition (ii) is satisfied. This completes the proof of the Lemma 5.2.1.

It is obvious from Definition 5.1.2 that the classes $\mathcal{TM}^*, \mathcal{TG}, \mathcal{TM}$ and $\mathcal{T}(\mathcal{MG})$ are invariant under transformations $\mathcal{T}_H$ induced by general vector fields H. In the next lemma we show that the classes $\mathcal{M}^*, \mathcal{G}, \mathcal{M}$ and $\mathcal{MG}$ are invariant under transformations $\mathcal{T}_A$ induced by a linear, bijective mapping $A : \mathbb{R}^N \to \mathbb{R}^N$. This result allows us to normalize transformation fields H by multiplication with an invertible matrix:

Lemma 5.2.2 *Let A be an invertible $N \times N$-matrix and let the transformation field $H : \mathbb{R}^N \to \mathbb{R}^N$ be given by the linear mapping $H(z) = Az$. Then $\mathcal{M}^*, \mathcal{G}, \mathcal{M}$ and $\mathcal{MG}$ are invariant under $\mathcal{T}_A$, hence $\mathcal{T}_A(\mathcal{M}^*) = \mathcal{M}^*, \mathcal{T}_A(\mathcal{G}) = \mathcal{G}, \mathcal{T}_A(\mathcal{M}) = \mathcal{M}$ and $\mathcal{T}_A(\mathcal{MG}) = \mathcal{MG}$. If $(f, B) \in \mathcal{M}^*$ with*

$$f(\varepsilon, z) = g(-\rho \nabla_z \psi(\varepsilon, z)),$$

where

$$\rho\psi(\varepsilon, z) = \frac{1}{2}[D(\varepsilon - Bz)] \cdot (\varepsilon - Bz) + \frac{1}{2}(Lz) \cdot z$$

is a positive semi-definite quadratic form, then the transformed function is given by

$$f_A(\varepsilon, h) = g_A(-\rho \nabla_h \psi_A(\varepsilon, h)) \tag{5.2.6}$$

with the vector field g_A and the positive semi-definite quadratic form ψ_A defined by

$$g_A = A \circ g \circ A^T$$

$$\rho\psi_A(\varepsilon, h) = \frac{1}{2}[D(\varepsilon - BA^{-1}h)] \cdot (\varepsilon - BA^{-1}h) + \frac{1}{2}[(A^{-1})^T LA^{-1}h] \cdot h. \tag{5.2.7}$$

If ψ is positive definite, then ψ_A is also positive definite. If g is monotone, then g_A is also monotone, and if $g(z) \cdot z \geq 0$ holds for all $z \in D(g)$, then $g_A(h) \cdot h \geq 0$ is also satisfied for all $h \in D(g_A)$. If $g = \nabla\chi$ for a function $\chi : \mathbb{R}^N \to \mathbb{R}$, then

$$g_A = \nabla\chi_A$$

where $\chi_A = \chi \circ A^T$. If χ is a convex function then so is χ_A.

Proof. To prove that $\mathcal{T}_A(\mathcal{M}^*) = \mathcal{M}^*$, it suffices to show that $(f_A, BA^{-1}) \in \mathcal{M}^*$, and to prove this relation, it suffices to show that (5.2.6) holds, that ψ_A defined in (5.2.7) is positive semi-definite, and that the symmetric linear mapping M_A defined by $M_A h = \rho \nabla_h \psi_A(0, h)$ is positive definite.

We first study ψ_A and use that (5.2.7) yields

$$\rho\psi_A(\varepsilon, h) = \frac{1}{2}[D(\varepsilon - BA^{-1}h)] \cdot (\varepsilon - BA^{-1}h) + \frac{1}{2}(LA^{-1}h) \cdot (A^{-1}h)$$

$$= \rho\psi(\varepsilon, A^{-1}h). \tag{5.2.8}$$

Therefore ψ_A is positive semi-definite if and only if ψ is positive semi-definite. Next we use that the derivative of the linear mapping $H^{-1} = A^{-1}$ coincides with A^{-1}. Since we consider gradients of real-valued functions as column vectors, we compute from the last equation that

$$\nabla_h \psi_A(\varepsilon, h) = \nabla_h \psi(\varepsilon, A^{-1}h) = [(\nabla_z \psi)^T(\varepsilon, A^{-1}h)A^{-1}]^T \qquad (5.2.9)$$
$$= (A^{-1})^T \nabla_z \psi(\varepsilon, A^{-1}h),$$

whence

$$M_A h = \rho \nabla_h \psi_A(0, h) = \rho (A^{-1})^T \nabla_z \psi(0, A^{-1}h) = (A^{-1})^T M A^{-1} h,$$

with the mapping M defined by $Mh = \rho \nabla_z \psi(0, h)$. This is positive definite by definition of the class $\mathcal{M}^*$. The last equation thus shows that M_A is also positive definite.

To prove that (5.2.6) holds, note that for the transformed function Lemma 5.1.1 yields

$$f_A(\varepsilon, h) = H'\big(H^{-1}(h)\big) f(\varepsilon, H^{-1}(h)) = A f(\varepsilon, A^{-1}h)$$
$$= Ag(-\rho \nabla_z \psi(\varepsilon, A^{-1}h)) = Ag(-\rho A^T \nabla_h \psi_A(\varepsilon, h)) = g_A(-\rho \nabla_h \psi_A(\varepsilon, h)),$$

where we used (5.2.9) to obtain the fourth equality. This shows that $\mathcal{T}_A(\mathcal{M}^*) = \mathcal{M}^*$ and proves the statements of the lemma about the form of the transformed equation.

To prove the remaining statements, note that (5.2.8) shows that ψ_A is positive definite if and only if ψ is positive definite. Since A is linear, it is obvious that $g_A = A \circ g \circ A^T$ is monotone if and only if g is monotone. Moreover, if $g(z) \cdot z \geq 0$ holds for all $z \in D(g)$, then

$$g_A(h) \cdot h = \big[A\big(g(A^T h)\big)\big] \cdot h = g(A^T h) \cdot A^T h \geq 0$$

for all $h \in D(g_A)$. Finally, if $g = \nabla \chi$, then

$$\nabla \chi_A(h) = \nabla_h \chi(A^T h) = [(\nabla \chi)^T(A^T h)A^T]^T$$
$$= A\nabla \chi(A^T h) = (A \circ g \circ A^T)(h) = g_A(h).$$

It follows from the linearity of the mapping A that if χ is a convex function, then so is $\chi_A = \chi \circ A^T$. This proves the lemma.

A pair (f, B) can be transformed to monotone type if it can be transformed to pre-monotone type and if for the transformed pair $(f_H, BH^{-1}) = (g(-\rho \nabla_h \psi), BH^{-1})$ the vector field g is monotone and satisfies $g(z) \cdot z \geq 0$. It is therefore of interest to know for a pair $(f, B) \in \mathcal{T}\mathcal{M}^*$ and for a vector field H which transforms this pair to the class $\mathcal{M}^*$, when the vector field g will be monotone. In the next lemma we state a condition for the monotonicity of g which can be formulated without knowing g. This condition can be used to determine whether a pair (f, B) can be transformed to monotone type, and thus to determine whether it belongs to the class $\mathcal{T}\mathcal{M}$. It can also be used to determine a transformation field H which performs this transformation.

If (f, B) is transformed to pre-monotone type by H, then (5.1.11) and therefore (5.2.3) hold. Since the mapping M and the quadratic form ψ from Corollary 5.1.3 satisfy $Mh = \rho \nabla_h \psi(0, h)$, we can write (5.2.3) in the form

$$H'(z) f(0, z) = g(-MH(z)).$$

Differentiation of this equation yields

$$\nabla_z[H'(z) f(0, z)] = H'(z)\nabla_z f(0, z) + H''(z) : f(0, z)$$
$$= \nabla g(-MH(z))(-M) H'(z),$$

with the tensor $H''(z)$ defined in Lemma 5.2.1. The last equation implies that

$$(\nabla g)(-MH(z)) = -H'(z)\nabla_z f(0, z) H'(z)^{-1}M^{-1}$$
$$- H''(z) : f(0, z) H'(z)^{-1}M^{-1}. \tag{5.2.10}$$

From this formula we obtain the following result:

Lemma 5.2.3 *If the vector field g in (5.2.3) is monotone, then the matrix*

$$H'(z)\nabla_z f(0, z) H'(z)^{-1}M^{-1} + H''(z) : f(0, z) H'(z)^{-1}M^{-1} \tag{5.2.11}$$

is negative semi-definite for all $z \in D_0(f)$. Conversely, if the matrix (5.2.11) is negative semi-definite for all $z \in D_0(f)$ and if in addition the set $H(D_0(f)) \subseteq \mathbb{R}^N$ is convex, then g in (5.2.3) is monotone on the convex set $-MH(D_0(f))$.

Proof. In the proof of Lemma 3.4.1 we showed that if the vector field $g : D(g) \subseteq \mathbb{R}^N \to \mathbb{R}^N$ is monotone, then the matrix $\nabla g(z)$ is positive semi-definite for all $z \in D(g)$, and conversely, if $\nabla g(z)$ is positive semi-definite for all z from the convex domain $D(g)$, then g is monotone. The statement of the lemma is an immediate consequence of this fact and of (5.2.10), since the image $-MH(D_0(f))$ of the convex set $H(D_0(f))$ under the linear map $-M$ is convex.

5.3   The Class of Constitutive Equations Transformable to Monotone Type

We summarize the preceding results and extend them slightly:

Theorem 5.3.1 *Let the pair (f, B) consist of the function f and the linear mapping B from the right hand side of the constitutive equations (5.1.2), (5.1.3).*

(I) *If a symmetric, positive definite $N \times N$-matrix M and a transformation field $H : \mathbb{R}^N \to \mathbb{R}^N$ exist satisfying the conditions*

(i) $BH^{-1} : \mathbb{R}^N \to S^3$ *is a linear mapping,*

(ii) $L = M - (BH^{-1})^T D(BH^{-1})$ *is positive semi-definite,*

(iii) $H'(z) f(\varepsilon, z)$ *is constant on every set $V(c, M, H) \cap D(f)$, where*

$$V(c, M, H) = \{(\varepsilon, z) \in S^3 \times \mathbb{R}^N \mid -\overline{M}\varepsilon + MH(z) = c\}$$

with $\overline{M} = (BH^{-1})^T D$,

(iv) *For every c the submanifold $V(c, M, H)$ intersects $\{0\} \times D_0(f)$ whenever it intersects $D(f)$,*

then $(f, B) \in \mathcal{TM}^*$ and the transformed equations of pre-monotone type have the form

$$T = D(\varepsilon - BH^{-1}(h)) \tag{5.3.1}$$

$$h_t = f_H(\varepsilon, h) = g(-\rho\nabla_h\psi(\varepsilon, h)). \tag{5.3.2}$$

Here the positive semi-definite quadratic form ψ and the vector field g are defined by

$$\rho\psi(\varepsilon, h) = \frac{1}{2}[D(\varepsilon - BH^{-1}(h))] \cdot (\varepsilon - BH^{-1}(h)) + \frac{1}{2}(Lh) \cdot h \tag{5.3.3}$$

$$g(h) = H'(H^{-1}(-M^{-1}h)) f(0, H^{-1}(-M^{-1}h)). \tag{5.3.4}$$

If in addition M, H and f satisfy

(v) *The $N \times N$-matrix*

$$H'(z)\nabla_z f(0, z) H'(z)^{-1} M^{-1} + H''(z) : f(0, z) H'(z)^{-1} M^{-1}$$

   is negative semi-definite for all $z \in D_0(f)$,

(vi) *$(H'(z) f(0, z)) \cdot (MH(z)) \leq 0$ for all $z \in D_0(f)$,*

(vii) *The set $H(D_0(f))$ is convex,*

then $(f, B) \in \mathcal{TM}$, and the vector field g defined in (5.3.4) is monotone.

Conversely, assume that $(f, B) \in \mathcal{TM}^*$ and let H be a vector field, that transforms (f, B) to a pair (f_H, BH^{-1}) of pre-monotone type, where f_H has the form (5.3.2) with a suitable vector field g and a suitable positive semi-definite free energy ψ. Then H and the matrix M defined by $Mh = \rho\nabla_h\psi(0, h)$ satisfy the conditions (i) – (iii). If $(f, B) \in \mathcal{TM}$, then M and H also satisfy the conditions (v) and (vi).

(II) *If condition (iii) holds, then*

$$H'(z)[\nabla_\varepsilon f(\varepsilon, z) + \nabla_z f(\varepsilon, z)H'(z)^{-1} M^{-1}\overline{M}] \tag{5.3.5}$$
$$+ H''(z) : f(\varepsilon, z) H'(z)^{-1} M^{-1}\overline{M} = 0$$

is satisfied. Consequently, for $(f, B) \in \mathcal{TM}^*$ the transformation field H and the matrix M defined by $Mh = \rho\nabla_h\psi(0, h)$ satisfy (5.3.5). On the other hand, if (5.3.5) is satisfied and if every one of the sets $V(c, M, H) \cap D(f)$ is connected, then condition (iii) holds.

Proof. (I) Assume that H and M satisfy (i) – (iv). To show that $(f, B) \in \mathcal{TM}^*$, it suffices to prove that the transformed pair (f_H, BH^{-1}) is of pre-monotone type. By inspection of Definition 3.1.1, we see that this will be true if (5.3.2) holds with the quadratic form ψ defined in equation (5.3.3), if L in this equation is symmetric, positive semi-definite and if $L + (BH^{-1})^T D(BH^{-1})$ is positive definite. L has these properties because of condition (ii) and because we assumed that M is positive definite. To show that (5.3.2) holds with the function g defined in (5.3.4), note that from (5.3.3) we obtain $Mh = \rho\nabla_h\psi(0,h)$, which shows that equation (5.2.3) is equivalent to (5.3.4). Therefore the function g from (5.3.4) satisfies (5.2.3). Together with conditions (iii) and (iv) we obtain from Lemma 5.2.1 that (5.1.11) holds, which because of (5.1.7) is nothing but (5.3.2). If additionally conditions (v) – (vii) hold, then we infer from Lemma 5.2.3 and conditions (v), (vii) that g is monotone, and condition (vi) together with (5.3.4) yields that for all $h \in D(g) = -MHD_0(f)$

$$g(h) \cdot h = [H'(H^{-1}(-M^{-1}h))\, f(0, H^{-1}(-M^{-1}h))] \cdot h$$
$$= -(H'(z)\, f(0,z)) \cdot MH(z) \geq 0, \qquad (5.3.6)$$

where we have set $h = -MH(z)$. According to Definition 3.1.1 we thus obtain that (5.3.1) and (5.3.2) are of monotone type, hence $(f, B) \in \mathcal{TM}$.

Next we prove the converse. By definition, (f_H, BH^{-1}) is of pre-monotone type if BH^{-1} is linear and if equation (5.3.2) holds with a free energy of the form (5.3.3), for which L must be positive semi-definite, and for which M given by

$$Mh = \rho\nabla_h\psi(0,h) = [(BH^{-1})^T D(BH^{-1}) + L]h$$

must be positive definite. Therefore M and H satisfy conditions (i) and (ii). From (5.1.7) we see that (5.3.2) is equivalent to (5.1.11), and consequently Lemma 5.2.1 shows that M and H also satisfy condition (iii). If $(f, B) \in \mathcal{TM}$, then $(f_H, BH^{-1}) \in \mathcal{M}$, which implies that g in (5.3.2) is monotone and satisfies $g(z) \cdot z \geq 0$ for all $z \in D(g)$. From Lemma 5.2.3 we thus obtain that (v) holds, whereas (vi) is implied by (5.3.6).

The assertion (II) in the theorem is just a restatement of the assertion (II) in Lemma 5.2.1.

This theorem yields criteria for constitutive equations to belong to the class of equations that can be transformed to pre-monotone or to monotone type. Still, these criteria do not show how to construct M and H satisfying the conditions of this theorem.

5.4 The Class of Constitutive Equations Transformable to Gradient Type

We next study when a pair (f, B) is transformable to gradient type. It turns out that in this case the negativity condition (v) of the preceding theorem is replaced by an interesting system of non-linear first order partial differential equations which must be satisfied by M and H. In principal, this opens the possibility of constructing H as the solution of a partial differential equation.

Thus, let the pair (f, B) be given. We want to determine H such that the transformed function f_H satisfies

$$f_H(\varepsilon, h) = \nabla\chi(-\rho\nabla_h\psi(\varepsilon, h)) \qquad (5.4.1)$$

for all $(\varepsilon, h) \in D(f_H)$, for a suitable function $\chi : \mathbb{R}^N \to \mathbb{R}$ and for the positive semi-definite quadratic form ψ given by (5.1.12). Using (5.1.7) we see that (5.4.1) is fulfilled if and only if

$$H'(H^{-1}(h))\,f(\varepsilon, H^{-1}(h)) = \nabla\chi(-\rho\nabla_h\psi(\varepsilon, h)) \tag{5.4.2}$$
$$= \nabla\chi\big((BH^{-1})^T D\varepsilon - Mh\big)$$

holds for all $(\varepsilon, h) \in D(f_H)$, where we used (5.2.1) to get the second equality. We set $\varepsilon = 0$ in (5.4.2) and obtain that H, M and χ must satisfy the equation

$$H'(z)f(0, z) = \nabla\chi\big(-\rho\nabla_h\psi(0, H(z))\big) = \nabla\chi(-MH(z)) \tag{5.4.3}$$

for all $z \in D_0(f)$. This is a system of N partial differential equations for the N components of H and for χ, but it is of an unusual form, since on the right hand side it contains the composition of the unknown functions H and χ. To bring it into an ordinary form, we define the function $\varphi : \mathbb{R}^N \to \mathbb{R}$ by

$$\varphi = \chi \circ (-M) \circ H .$$

Since we consider gradients of real-valued functions as column vectors, the chain rule yields

$$(\nabla\varphi(z))^T = [\nabla\chi(-MH(z))]^T (-M \circ H)'(z) ,$$

hence

$$\nabla\varphi = -[(M \circ H)']^T (\nabla\chi \circ (-M) \circ H) .$$

Multiplication of (5.4.3) with $-[(M \circ H)']^T$ thus results in

$$-[(M \circ H)'(z)]^T\, H'(z)f(0, z) = -[(M \circ H)'(z)]^T \nabla\chi(-MH(z))$$
$$= \nabla\varphi(z) ,$$

or, equivalently,

$$- H'(z)^T M H'(z) f(0, z) = \nabla\varphi(z) \tag{5.4.4}$$

for all $z \in D_0(f)$. Here we used the symmetry of the linear mapping M. Using

$$H'(z)\, f(0, z) = (f(0, z) \cdot \nabla)\, H(z)$$

and

$$[(M \circ H)'(z)]^T = \big(\nabla(M_1 \cdot H(z)), \ldots, \nabla(M_N \cdot H(z))\big) ,$$

where $M_i = (M_{i1}, \ldots, M_{iN})$ denotes the i-th row of the $N \times N$-matrix M, we can also write (5.4.4) in the form

$$\sum_{i=1}^N (f(0, z) \cdot \nabla H_i(z))\, \nabla(M_i \cdot H(z)) + \nabla\varphi(z) = 0 . \tag{5.4.5}$$

(5.4.4) or (5.4.5) is a nonlinear system of N first order partial differential equations for the components $H_i : \mathbb{R}^N \to \mathbb{R}$, $i = 1, \ldots, N$, of the vector field H and for the real valued function $\varphi : \mathbb{R}^N \to \mathbb{R}$, and we showed that if H and χ satisfy (5.4.3), then H and φ satisfy (5.4.4). Going through the above computations in the opposite direction, one verifies immediately that if H and φ satisfy (5.4.4), then H and the function χ defined by

$$\chi = \varphi \circ H^{-1} \circ (-M)^{-1}$$

satisfy (5.4.3). We summarize these results:

Theorem 5.4.1 (I) *Assume that the vector field* $H : \mathbb{R}^N \to \mathbb{R}^N$ *transforms the constitutive equations* (5.1.2), (5.1.3) *to the equations*

$$T = D(\varepsilon - BH^{-1}(h)) \tag{5.4.6}$$
$$h_t = f_H(\varepsilon, h) = g(-\rho\nabla_h\psi(\varepsilon, h)) \tag{5.4.7}$$

of pre-monotone type, and let the symmetric, positive definite $N \times N$*-matrix* M *be defined by* $Mh = \rho\nabla_h\psi(0, h)$.

Then g *from* (5.4.7) *is a gradient field on* $-MH(D_0(f))$ *and therefore satisfies*

$$g(h) = \nabla\chi(h), \quad h \in -MH(D_0(f)) \tag{5.4.8}$$

for a suitable function $\chi : -MH(D_0(f)) \to \mathbb{R}$, *if and only if* H *and the function*

$$\varphi = \chi \circ (-M) \circ H : D_0(f) \to \mathbb{R}$$

satisfy the system of N *first order partial differential equations*

$$-H'(z)^T M H'(z) f(0, z) = \nabla\varphi(z) \tag{5.4.9}$$

for all $z \in D_0(f)$.

(II) *If the pair* (f, B) *belongs to the class* $\mathcal{TG}$ *and if* H *is a vector field, which transforms the pair* (f, B) *to gradient type, then* H *transforms this pair to pre-monotone type and, together with a suitable function* $\varphi : D_0(f) \to \mathbb{R}$, *satisfies* (5.4.9). *Conversely, if* H *is a vector field which transforms the pair* (f, B) *to pre-monotone type, if* (5.4.9) *is satisfied and condition* (iv) *of Theorem 5.3.1 holds, then* (f, B) *is transformed to gradient type by* H, *and consequently* (f, B) *belongs to the class* $\mathcal{TG}$.

Proof. (I) Before Theorem 5.4.1 we already proved that (5.4.9) is equivalent to (5.4.3) and together with (5.1.7) we see that this in turn is equivalent to

$$f_H(0, h) = H'(H^{-1}(h)) f(0, H^{-1}(h)) = \nabla\chi(-Mh) = \nabla\chi(-\rho\nabla_h\psi(0, h))$$

for all $h \in H(D_0(f))$. By comparison with (5.4.7), we see that this equation holds if and only if $g(h) = \nabla\chi(h)$ for all $h \in -MH(D_0(f))$. This proves (I).

(II) H transforms (f, B) to gradient type if and only if H transforms (f, B) to pre-monotone type and satisfies (5.4.2) for all $(\varepsilon, h) \in D(f_H)$. Since (5.4.2) implies (5.4.3), and since (5.4.3) and (5.4.9) are equivalent, we obtain one direction of statement (II). To obtain the converse, it suffices to show that when condition (iv) of Theorem 5.3.1 holds, then (5.4.3) implies (5.4.2). Now, by assumption, H transforms (f, B) to pre-monotone type, and thus satisfies not only condition (iv), but also condition (iii) of Theorem 5.3.1. These two conditions and (5.4.3) show that the assumptions of Lemma 5.2.1 are satisfied with $g = \nabla\chi$ in (5.2.3), hence (5.1.11) holds, which is (5.4.2). The proof is complete.

5.5 The Class of Constitutive Equations Transformable to Monotone-Gradient Type

Together with Theorem 5.3.1 the last theorem characterizes the class of those constitutive equations (5.1.2), (5.1.3), which can be transformed to gradient type. Of

course, the transformed equations need not be of monotone type, but it is an interesting observation that for many non-monotone constitutive equations from the
class $\mathcal{TG}$, transformation to gradient type automatically yields constitutive equations of monotone-gradient type, and hence of monotone type. In the next theorem
we give a criterion for this to happen, and thus, in fact, study the class $\mathcal{T(MG)}$.

To find such a criterion, it must be investigated, when the function χ from (5.4.8)
is convex. If the domain of χ is convex, then χ is convex if and only if the Hessian
matrix $\nabla^2\chi(h)$ is positive semi-definite. Therefore we must derive criteria for this
Hessian to be positive semi-definite. The criterion stated in the next theorem is
based on the symmetry of Hessian matrices, a property which, of course, a matrix
$\nabla g(h)$ to an arbitrary vector field g normally does not have.

To state the following theorem, we need some new notations. With a symmetric,
positive definite $N \times N$–matrix M a scalar product on $\mathbb{R}^N$ is defined by

$$(M^{-1}\zeta) \cdot \eta, \quad \zeta, \eta \in \mathbb{R}^N,$$

and

$$|\zeta|_{M^{-1}} = [(M^{-1}\zeta) \cdot \zeta]^{1/2}$$

is the norm associated to this scalar product. This norm induces the operator norm

$$\|A\|_{M^{-1}} = \sup \{|A\zeta|_{M^{-1}} \mid |\zeta|_{M^{-1}} \le 1\}$$

of a linear operator $A : \mathbb{R}^N \to \mathbb{R}^N$.

Theorem 5.5.1 *Assume that the constitutive equations (5.1.2), (5.1.3) are transformed by the vector field H into the equations*

$$T = D(\varepsilon - BH^{-1}(h)) \tag{5.5.1}$$
$$h_t = f_H(\varepsilon, h) = \nabla\chi(-\rho\nabla_h\psi(\varepsilon, h)) \tag{5.5.2}$$

*of gradient type, and let the symmetric, positive definite matrix M be defined by
$Mh = \rho\nabla_h\psi(0, h)$.*

(I) If the function χ in (5.5.2) is convex, then the matrix

$$- H'(z)\nabla_z f(0, z) H'(z)^{-1}M^{-1} - H''(z) : f(0, z) H'(z)^{-1} M^{-1} \tag{5.5.3}$$

is positive semi-definite for all $z \in D_0(f)$.

*Conversely, if the matrix (5.5.3) is positive semi-definite for all $z \in D_0(f)$ and if
the set $-MH(D_0(f))$ is convex, then χ is convex on this set. Consequently, if the
function χ also satisfies*

$$\nabla\chi(h) \cdot h \ge 0 \tag{5.5.4}$$

*for all $h \in D(\chi)$, then the constitutive equations (5.5.1), (5.5.2) are of monotone-
gradient type, and therefore (5.1.2), (5.1.3) belong to $\mathcal{T(MG)}$.*

*(II) Let $z \in D_0(f)$, let $\lambda_1, \ldots, \lambda_m \in \mathbb{C}$, $\lambda_i = \lambda_i(z)$, be the eigenvalues of the $N \times N$–
matrix $-\nabla_z f(0, z)$, and assume that all eigenvalues have non-negative real part.
With*

$$\nu = \nu(z) = \frac{1}{2N - 1} \min_{1 \le i \le m} |\lambda_i(z)| \tag{5.5.5}$$

suppose that

$$\|MH''(z) : f(0,z)\,H'(z)^{-1}M^{-1}\|_{M^{-1}} \leq \nu(z).\tag{5.5.6}$$

Then the matrix (5.5.3) *is positive semi-definite. Consequently, if χ satisfies* (5.5.4), *if* (5.5.6) *holds for all $z \in D_0(f)$, and if $-MH(D_0(f))$ is convex, then $(f,B) \in \mathcal{T}(\mathcal{MG})$.*

Proof. (I) Insertion of $g = \nabla\chi$ into (5.2.10) yields

$$(\nabla^2\chi)(-MH(z)) = -H'(z)\nabla_z f(0,z)\,H'(z)^{-1}M^{-1}$$
$$-H''(z) : f(0,z)\,H'(z)^{-1}\,M^{-1},\tag{5.5.7}$$

which means that the matrix in (5.5.3) is positive semi-definite for all $z \in D_0(f)$ if and only if $\nabla^2\chi(h)$ is positive semi-definite for all $h \in -MH(D_0(f))$. Therefore part (I) of the theorem follows immediately from the fact that a twice continuously differentiable function χ is convex if and only if the domain is convex and $\nabla^2\chi$ is positive semi-definite on the domain.

(II) To prove the second part, we split the expression (5.5.3) into the parts

$$X_1 = -H'(z)\nabla_z f(0,z)H'(z)^{-1}M^{-1}$$
$$X_2 = -H''(z) : f(0,z)\,H'(z)^{-1}M^{-1}.$$

Equation (5.5.7) yields

$$\nabla^2\chi(-MH(z)) = X_1 + X_2.$$

The $N \times N$-matrix $\nabla^2\chi(h)$ and therefore $X_1 + X_2$ is symmetric. This implies that the linear mapping $MX_1 + MX_2$ is symmetric with respect to the scalar product $(M^{-1}\zeta)\cdot\eta$ on $\mathbb{R}^N$, and consequently its spectrum

$$\Sigma = \{\omega_1, \ldots, \omega_N\}$$

consists of eigenvalues ω_i which are all real. Here we count the eigenvalues according to multiplicity. For $r > 0$ let

$$K_r(\Sigma) = \{\xi \in \mathbb{C} \mid \mathrm{dist}(\xi, \Sigma) \leq r\}$$

be the union of the closed circles with radius r and centers $\omega_1, \ldots, \omega_N$. We now use the following statement:

(S) *Let $r \geq 0$. If $Y : \mathbb{R}^N \to \mathbb{R}^N$ is a linear operator with $\|Y\|_{M^{-1}} \leq r$, then every connected component of the set $K_r(\Sigma)$ contains at least one eigenvalue of the operator $MX_1 + MX_2 + Y$.*

Before proving this statement, we first show that the assertion of the lemma is a consequence of (S): Let $r \geq 0$. If ω_i is an eigenvalue of $MX_1 + MX_2$, then the connected component of $K_r(\Sigma)$ containing ω_i is contained in the closed circle of radius $2r(N-1) + r$ and center ω_i, since $K_r(\Sigma)$ is the union of N circles of radius r. If we choose $Y = -MX_2$ and $r = \nu$, it thus follows from (S) and from (5.5.6) that the circle around ω_i with radius

$$2\nu(N-1) + \nu = (2N-1)\nu$$

contains an eigenvalue of $MX_1 + MX_2 + Y = MX_1$. The eigenvalues of MX_1 coincide with the eigenvalues of $-\nabla_z f(0, z)$, since a vector $\zeta \in \mathbb{C}^N$, $\zeta \neq 0$, and a number $\lambda \in \mathbb{C}$ satisfy the equation

$$MX_1 \zeta = -MH'(z)\nabla_z f(0, z)H'(z)^{-1}M^{-1}\zeta = \lambda\zeta$$

if and only if λ and $\eta = H'(z)^{-1}M^{-1}\zeta$ satisfy the equation

$$-\nabla_z f(0, z)\eta = \lambda\eta.$$

This is seen by multiplying the first equation by $H'(z)^{-1}M^{-1}$. Therefore the circle with center ω_i and radius $(2N - 1)\nu$ contains an element of the set $\{\lambda_1, \ldots, \lambda_m\}$ of eigenvalues of $-\nabla_z f(0, z)$. Let this element be λ_ℓ. Then, vice versa, ω_i is contained in the closed circle with center λ_ℓ and radius $(2N - 1)\nu$. According to (5.5.5), this radius satisfies

$$(2N - 1)\nu = \min_{1 \leq j \leq m} |\lambda_j|.$$

Therefore it is not larger than $|\lambda_\ell|$. Since ω_i is real and since $\mathrm{Re}\,\lambda_\ell \geq 0$, this can only be if $\omega_i \geq 0$. Because ω_i is an arbitrary eigenvalue of $MX_1 + MX_2$, it follows that all eigenvalues of the mapping $MX_1 + MX_2$ are non-negative, and together with the symmetry with respect to the scalar product $(M^{-1}\zeta) \cdot \eta$, this implies that this mapping is positive semi-definite with respect to this scalar product. This yields

$$[(X_1 + X_2)\zeta] \cdot \zeta = [M^{-1}(MX_1 + MX_2)\zeta] \cdot \zeta \geq 0$$

for all $\zeta \in \mathbb{R}^N$. Consequently, the matrix (5.5.3), which is equal to $X_1 + X_2$, is positive semi-definite, and the assertion of the lemma holds.

It remains to prove statement (S). This statement follows from well known results of spectral theory. Let $\xi \in \mathbb{C}\backslash\Sigma$ be an element from the resolvent set of $MX_1 + MX_2$ with

$$\|Y\|_{M^{-1}} < \frac{1}{\|(MX_1 + MX_2 - \xi I)^{-1}\|_{M^{-1}}}. \tag{5.5.8}$$

Then ξ belongs to the resolvent set of $MX_1 + MX_2 + Y$, [61, p. 585]. Since $MX_1 + MX_2$ is symmetric, we have

$$\|(MX_1 + MX_2 - \xi I)^{-1}\|_{M^{-1}} = \frac{1}{\mathrm{dist}(\xi, \Sigma)}.$$

From this equation and from (5.5.8) we obtain that ξ belongs to the resolvent set of $MX_1 + MX_2 + Y$, if

$$\|Y\|_{M^{-1}} < \mathrm{dist}(\xi, \Sigma).$$

This means that the spectrum of $MX_1 + MX_2 + Y$ is contained in the set

$$\{\xi \in \mathbb{C} \mid \mathrm{dist}(\xi, \Sigma) \leq \|Y\|_{M^{-1}}\},$$

which is a subset of $K_r(\Sigma)$ if $\|Y\|_{M^{-1}} \leq r$. Therefore all eigenvalues of $MX_1 + MX_2 + Y$ are contained in $K_r(\Sigma)$, and it remains to show that every connected component of $K_r(\Sigma)$ contains at least one eigenvalue.

For $t \in [0,1]$ let $Z(t) = MX_1 + MX_2 + tY$. Then there exist continuous functions $\mu_i : [0,1] \to \mathbb{C}$, $i = 1, \ldots, N$, such that the set of eigenvalues of $Z(t)$ is equal to $\{\mu_1(t), \ldots, \mu_N(t)\}$, cf. [109, p. 109]. This implies that $\{\mu_1(t), \ldots, \mu_N(t)\} \subseteq K_r(\Sigma)$, since $\|tY\|_{M^{-1}} \leq \|Y\|_{M^{-1}} \leq r$, and it also implies that to every ω_i there exists at least one μ_j with $\mu_j(0) = \omega_i$. Because μ_j is continuous, it follows that $\mu_j(1)$ belongs to the same connected component of $K_r(\Sigma)$ as ω_i. Now, since $K_r(\Sigma)$ is a union of circles with centers $\omega_1, \ldots, \omega_N$, we obtain that every connected component contains at least one ω_i. Hence it also contains $\mu_j(1)$, which is an eigenvalue of $Z(1) = MX_1 + MX_2 + Y$. This proves statement (S).

Chapter 6

Classification Conditions

In the preceding chapter we showed that if constitutive equations can be transformed to gradient type by a vector field H, then H must satisfy the system of partial differential equations (5.4.9) and also conditions (i) - (iii) of Theorem 5.3.1. Condition (iii) is equivalent to the system of partial differential equations (5.3.5), and together (5.3.5) and (5.4.9) form an overdetermined system of partial differential equations for H. Solutions H of this system satisfying also conditions (i) and (ii) exist only if the coefficients in this system fulfill suitable compatibility conditions. Because these coefficients are essentially given by the function f from the constitutive equations and its derivatives, this means that f must satisfy compatibility conditions. In turn this means that these compatibility conditions are classification conditions for constitutive equations: a pair (f, B) belongs to the class $\mathcal{TG}$ introduced in Definition 5.1.2 if and only if it satisfies the compatibility conditions. Of course, these classification conditions are equivalent to the conditions given in Theorems 5.3.1 and 5.4.1, but one can hope that the classification conditions are more explicit and can be tested more easily than the conditions of these theorems. In particular, it seems probable that simpler conditions can be obtained if one starts from the already restricted class of constitutive equations of pre-monotone type and asks what classification conditions must be satisfied in order that these equations can be transformed to gradient type.

This is an interesting question, since, as the examples in Sections 2 and 3 show, on the one hand almost all constitutive equations used in solid mechanics are of pre-monotone type, and on the other hand the class $\mathcal{MG}$ of monotone-gradient type is a subclass of $\mathcal{G}$. Furthermore, transformation of constitutive equations to the class $\mathcal{G}$ often automatically yields equations of the class $\mathcal{MG}$. This was discussed at the end of the preceding chapter and is also illustrated by the examples of Chapter 8. This observation is of special importance for rate independent constitutive equations, which are studied in Chapter 7, since rate independent constitutive equations of monotone type are necessarily of monotone-gradient type, and transformation of rate independent equations to monotone type therefore always requires them to be transformed to gradient type.

In this chapter we derive such classification conditions, starting from a class of constitutive equations which is even more general then the class $\mathcal{M}^*$ of pre-monotone type. Since $\mathcal{G}$ is a subclass of $\mathcal{M}^*$, in Section 6.1 we first investigate under what conditions constitutive equations of this more general class can be transformed to monotone type. One of the two main results of this section is, however, that in most cases this is only possible if the constitutive equation to be transformed is already of pre-monotone type. The other main result is the determination of the class of transformation fields H which transform $\mathcal{M}^*$ into itself, hence for which $T_H(\mathcal{M}^*) \subseteq \mathcal{M}^*$ holds. Both of these results are formulated in Corollary 6.1.4. In Section 6.2 we subsequently investigate which constitutive equations of $\mathcal{M}^*$ can be transformed to gradient type by transformation fields of this special form.

6.1 Transformations which Leave the Class of Pre-Monotone Equations Invariant

We study the transformation to gradient type of the constititive equations in the system

$$\rho u_{tt} = \mathrm{div}_x\, T \tag{6.1.1}$$

$$T = D(\varepsilon(\nabla_x u) - Bz) \tag{6.1.2}$$

$$z_t = f(\varepsilon(\nabla_x u), z) \tag{6.1.3}$$

with

$$f(\varepsilon, z) = g(\overline{L}\varepsilon - \hat{L}z). \tag{6.1.4}$$

Here $g : D(g) \subseteq \mathbb{R}^N \to \mathbb{R}^N$ is a given function,

$$B : \mathbb{R}^N \to \mathcal{S}^3,\ \overline{L} : \mathcal{S}^3 \to \mathbb{R}^N$$

are linear mappings, and

$$\hat{L} : \mathbb{R}^N \to \mathbb{R}^N$$

is an invertible linear mapping. The domain of f is given by

$$D(f) = \{(\varepsilon, z) \in \mathcal{S}^3 \times \mathbb{R}^N \mid \overline{L}\varepsilon - \hat{L}z \in D(g)\}. \tag{6.1.5}$$

The class of these constitutive equations is larger than the class $\mathcal{M}^*$, since the conditions for the linear mappings $\overline{L}$ and $\hat{L}$ are less restrictive than the corresponding conditions required in the pre-monotone case. We first study the restrictions imposed by the conditions of Theorem 5.3.1 on vector fields $H : \mathbb{R}^N \to \mathbb{R}^N$ that transform this system to a system of pre-monotone type given by

$$\rho u_{tt} = \mathrm{div}_x\, T$$

$$T = D(\varepsilon(\nabla_x u) - BH^{-1}h) \tag{6.1.6}$$

$$h_t = f_H(\varepsilon(\nabla_x u), h) \tag{6.1.7}$$

with

$$f_H(\varepsilon, h) = g_H(-\rho\nabla_h \psi(\varepsilon, h)) = g_H(\overline{M}_H \varepsilon - M_H h). \tag{6.1.8}$$

Here, according to Definition 3.1.1, M_H is a symmetric, positive definite $N \times N$-matrix such that

$$L_H = M_H - (BH^{-1})^T D(BH^{-1})$$

is positive semi-definite. The free energy ψ is given by

$$\rho\psi(\varepsilon, h) = \frac{1}{2}[D(\varepsilon - BH^{-1}h)] \cdot (\varepsilon - BH^{-1}h) + \frac{1}{2}(L_H h) \cdot h,$$

and consequently the linear mapping $\overline{M}_H$ is equal to

$$\overline{M}_H = (BH^{-1})^T D : \mathcal{S}^3 \to \mathbb{R}^N. \tag{6.1.9}$$

Our goal is to determine H such that the vector field g_H is the gradient of a convex function. As preparation of the investigations in this section we need the following technical result:

Lemma 6.1.1 *Let $A : \mathbb{R}^N \to \mathcal{S}^3$ be a linear mapping and let $M : \mathbb{R}^N \to \mathbb{R}^N$ be a symmetric, positive definite linear mapping. Then the kernel*

$$\ker(A) = \{z \in \mathbb{R}^N \mid Az = 0\}$$

of A and the range

$$R(M^{-1}A^T D) = \{M^{-1}A^T D\varepsilon \in \mathbb{R}^N \mid \varepsilon \in \mathcal{S}^3\}$$

of the linear mapping $M^{-1}A^T D : \mathcal{S}^3 \to \mathbb{R}^N$ are complementary subspaces of $\mathbb{R}^N$, which means that

$$\ker(A) \cap R(M^{-1}A^T D) = \{0\}, \quad \ker(A) + R(M^{-1}A^T D) = \mathbb{R}^N.$$

Proof. Since D is symmetric, it follows that $A^T D = (DA)^T$, hence

$$R(M^{-1}A^T D) = M^{-1}R(A^T D) = M^{-1}R((DA)^T). \tag{6.1.10}$$

By a well known result from linear algebra, the range of $(DA)^T$ and the kernel $\ker(DA)$ are orthogonal spaces:

$$R((DA)^T) = [\ker(DA)]^\perp.$$

Since D is positive definite and consequently invertible, $\ker(A) = \ker(DA)$, whence

$$R((DA)^T) = [\ker(A)]^\perp. \tag{6.1.11}$$

As orthogonal spaces $R((DA)^T)$ and $\ker(A)$ are therefore complementary spaces. Because (6.1.10) shows that $R(M^{-1}A^T D)$ is isomorphic to $R((DA)^T)$, we thus obtain that $R(M^{-1}A^T D)$ and $\ker(A)$ will be complementary subspaces, if

$$R(M^{-1}A^T D) \cap \ker(A) = \{0\}. \tag{6.1.12}$$

To prove (6.1.12), let $z \in R(M^{-1}A^T D) \cap \ker(A)$. From (6.1.10) we then obtain

$$Mz \in R((DA)^T).$$

Since $z \in \ker(A)$, we conclude from (6.1.11) that $(Mz) \cdot z = 0$, whence $z = 0$, since M is assumed to be positive definite. This proves (6.1.8) and completes the proof of the lemma.

For complementary subspaces U and V of $\mathbb{R}^N$ we can define a projection $P : \mathbb{R}^N \to \mathbb{R}^N$ with

$$P(\mathbb{R}^N) = V, \quad \ker(P) = U.$$

Then $Q = I - P$ is also a projection with

$$Q(\mathbb{R}^N) = U, \quad \ker(Q) = V.$$

P is called the projection of $\mathbb{R}^N$ onto V along U. Accordingly, Q is the projection onto U along V.

We first introduce a class of vector fields H which always satisfy the restrictions imposed by Theorem 5.3.1:

Theorem 6.1.2 *Let the pair (f, B) consist of the function f and the linear mapping from the constitutive equations (6.1.2) – (6.1.4), and let $\overline{L}, \hat{L}$ be the linear mappings from (6.1.4).*
Assume that a symmetric, positive definite $N \times N$–matrix M exists with

(i) $M^{-1}\overline{M} = \hat{L}^{-1}\overline{L}$,   *where* $\overline{M} = B^T D$,   (6.1.13)

(ii) $L = M - B^T D B$ *is positive semi-definite.*

Then $(f, B) \in \mathcal{M}^*$. Moreover, also the pair obtained by transformation of (f, B) with a vector field H of the form

$$H(z) = Pz + \tilde{H}(Qz) \tag{6.1.14}$$

belongs to $\mathcal{M}^*$. Here P is the projection along $\ker(B)$ onto the range

$$R(\hat{L}^{-1}\overline{L}) = \{\hat{L}^{-1}\overline{L}\varepsilon \in \mathbb{R}^N \mid \varepsilon \in \mathcal{S}^3\}$$

of the linear map $\hat{L}^{-1}\overline{L} : \mathcal{S}^3 \to \mathbb{R}^N$, $Q = I - P$ is the projection along $R(\hat{L}^{-1}\overline{L})$ onto $\ker(B)$, and $\tilde{H} : \ker(B) \to \mathbb{R}^N$ is a function with $\tilde{H}(\ker(B)) = \ker(B)$, such that

$$\tilde{H} : \ker(B) \to \ker(B)$$

is bijective. The transformed constitutive equations are of the form

$$T = D(\varepsilon - Bh) \tag{6.1.15}$$

$$h_t = g_H(-\rho\nabla_h\psi(\varepsilon, h)) = g_H(\overline{M}\varepsilon - Mh) \tag{6.1.16}$$

with

$$\rho\psi(\varepsilon, h) = \frac{1}{2}[D(\varepsilon - Bh)] \cdot (\varepsilon - Bh) + \frac{1}{2}(Lh) \cdot h, \tag{6.1.17}$$

and with a suitable vector field $g_H : D(g_H) = -MH(-\hat{L}^{-1}D(g)) \subseteq \mathbb{R}^N \to \mathbb{R}^N$.

The projectors P and Q can be defined, since $\ker(B)$ and $R(\hat{L}^{-1}\overline{L})$ are complementary subspaces of $\mathbb{R}^N$. This follows immediately from Lemma 6.1.1 with $A = B$, since (6.1.13) implies

$$\hat{L}^{-1}\overline{L} = M^{-1}\overline{M} = M^{-1}B^T D.$$

Proof. We show that M and all vector fields of the form (6.1.14) satisfy the conditions (i) – (iv) of Theorem 5.3.1, and in particular, that $BH^{-1} = B$. Then Theorem

5.3.1 implies that the transformed pair $(f_H, BH^{-1}) = (f_H, B)$ belongs to $\mathcal{M}^*$ and that the equations obtained by transformation of (6.1.2)–(6.1.4) with H have the form (6.1.15), (6.1.16) with ψ defined in (6.1.17). The domain of the function g_H in (6.1.16) is obtained from the equation

$$g_H(\overline{M}\varepsilon - Mh) = f_H(\varepsilon, h)$$
$$= H'(H^{-1}(h))f(\varepsilon, H^{-1}(h)) = H'(H^{-1}(h))g(\overline{L}\varepsilon - \hat{L}H^{-1}(h)),$$

which holds for all (ε, h) with $(\varepsilon, H^{-1}(h)) \in D(f)$. This implies $g_H(-Mh) = H'(H^{-1}(h))g(-\hat{L}H^{-1}(h))$, from which we conclude that $-Mh \in D(g_H)$ if and only if $-\hat{L}H^{-1}(h) \in D(g)$. Hence $D(g_H) = -MH(-\hat{L}^{-1}D(g))$.

To prove that the conditions (i) – (iv) of Theorem 5.3.1 are satisfied, note that if, by a slight abuse of notation, we denote the inverse of $\tilde{H} : \ker(B) \to \ker(B)$ by $\tilde{H}^{-1}$, then

$$H^{-1}(h) = Ph + \tilde{H}^{-1}(Qh), \tag{6.1.18}$$

since (6.1.14) and the relation $PQ = QP = 0$ satisfied by the projections P and Q yield that

$$H^{-1}(H(z)) = P(Pz + \tilde{H}(Qz)) + \tilde{H}^{-1}\big(Q(Pz + \tilde{H}(Qz))\big)$$
$$= Pz + PQ\tilde{H}(Qz) + \tilde{H}^{-1}(Q\tilde{H}(Qz))$$
$$= Pz + \tilde{H}^{-1}(\tilde{H}(Qz))$$
$$= Pz + Qz = z.$$

In the same way

$$H(H^{-1}(h)) = P(Ph + \tilde{H}^{-1}(Qh)) + \tilde{H}\big(Q(Ph + \tilde{H}^{-1}(Qh))\big) = h.$$

Thus

$$BH^{-1}(h) = B(Ph + \tilde{H}^{-1}(Qh))$$
$$= BPh + B\tilde{H}^{-1}(Qh)$$
$$= BPh = BPh + BQh$$
$$= B(P + Q)h = Bh,$$

since $\tilde{H}^{-1}(Qh), Qh \in \ker(B)$. This proves that $BH^{-1} = B$, and incidentally, that condition (i) of Theorem 5.3.1 is satisfied. Condition (ii) of Theorem 5.3.1 is equivalent to assumption (ii) of the present theorem.

To prove that conditions (iii) and (iv) of Theorem 5.3.1 are satisfied, we first determine the set $V(c, M, H)$. By definition, this set consists of all $(\varepsilon, z) \in \mathcal{S}^3 \times \mathbb{R}^N$ with

$$-\overline{M}\varepsilon + MH(z) = c.$$

Together with (6.1.13) and (6.1.18) we thus obtain for $(\varepsilon, z) \in V(c, M, H)$ that

$$z = H^{-1}(M^{-1}\overline{M}\varepsilon + M^{-1}c) = H^{-1}(\hat{L}^{-1}\overline{L}\varepsilon + M^{-1}c)$$
$$= P(\hat{L}^{-1}\overline{L}\varepsilon + M^{-1}c) + \tilde{H}^{-1}(Q\hat{L}^{-1}\overline{L}\varepsilon + QM^{-1}c) \tag{6.1.19}$$
$$= \hat{L}^{-1}\overline{L}\varepsilon + PM^{-1}c + \tilde{H}^{-1}(QM^{-1}c),$$

where we used that P projects onto $R(\hat{L}^{-1}\overline{L})$ and that $\ker(Q) = R(\hat{L}^{-1}\overline{L})$. From this equation and from (6.1.4) we infer that

$$f(\varepsilon, z) = g(\overline{L}\varepsilon - \hat{L}z) = g\left(-\hat{L}(PM^{-1}c + \tilde{H}^{-1}(QM^{-1}c))\right), \qquad (6.1.20)$$

and so $f(\varepsilon, z)$ is constant for all $(\varepsilon, z) \in V(c, M, H)$. Moreover, since (6.1.14) implies

$$H'(z) = P + \tilde{H}'(Qz)Q,$$

equation (6.1.19) yields

$$\begin{aligned}
H'(z) &= P + \tilde{H}'(Q\hat{L}^{-1}\overline{L}\varepsilon + QPM^{-1}c + Q\tilde{H}^{-1}(QM^{-1}c))Q \\
&= P + \tilde{H}'(\tilde{H}^{-1}(QM^{-1}c))Q = \text{const}
\end{aligned} \qquad (6.1.21)$$

on $V(c, M, H)$, where we again used that $Q\hat{L}^{-1}\overline{L}\varepsilon = 0$. Equations (6.1.20) and (6.1.21) together imply that $H'(z)f(\varepsilon, z)$ is constant on the set $V(c, M, H)$, which proves condition (iii) of Theorem 5.3.1. Finally, condition (iv) immediately results from the definition of $D(f)$ in (6.1.5), since (6.1.19) implies that $\overline{L}\varepsilon - \hat{L}z$ is constant on $V(c, M, H)$.

To complete the proof of Theorem 6.1.2, it remains to ensure that $(f, B) \in \mathcal{M}^*$. This is an immediate consequence of the other assertions of the theorem, which we already proved. Namely, by letting $\tilde{H}$ in (6.1.14) be the identity on $\ker(B)$, the vector field H becomes the identity on $\mathbb{R}^N$, and consequently the pair (f, B) itself must belong to $\mathcal{M}^*$. The proof of Theorem 6.1.2 is complete.

This theorem essentially shows that vector fields of the form (6.1.14) always transform the constitutive equations (6.1.2) – (6.1.4) to pre-monotone type. However, this is a restricted class of transformation fields, since they transform the subspaces $\ker(B)$ and $R(\hat{L}^{-1}\overline{L})$ into itself, and moreover, act as the identity on the latter subspace. For the transformed interior variables $h(x, t)$ this means that

$$Ph(x, t) = PH(z(x, t)) = Pz(x, t),$$

and consequently, some of the interior variables do not change under transformation with H. It is clear that for suitable functions g in (6.1.4) more general transformation fields are allowed. To take a trivial example, if the function g in (6.1.4) is constant, then all transformation fields are allowed for which BH^{-1} is linear. It is easy to construct vector fields that satisfy this condition, but are not of the form (6.1.14). However, in the next theorem we show that under a certain 'covering condition' for g, which essentially excludes that certain derivatives of g uniformly vanish on special affine submanifolds of the domain, all vector fields that transform the constitutive equations (6.1.2) – (6.1.4) to pre-monotone type can be normalized to have the form (6.1.14) on the set of all $z \in \mathbb{R}^N$, for which

$$D(f, z) = \{\varepsilon \in \mathcal{S}^3 \mid (\varepsilon, z) \in D(f)\}$$

is not empty. One cannot expect that such a strong restriction of $H(z)$ holds for values of z with $D(f, z) = \emptyset$, since this restriction is caused by conditions (i) and (iii) of Theorem 5.3.1, and (iii) is empty for such z. However, (5.1.7) shows that for the transformation of f only those values of $H(z)$ with $D(f, z) \neq \emptyset$ are important. This means that under the covering condition for g, all vector fields that transform these constitutive equations to pre-monotone type are essentially given by (6.1.14). We mention that the condition for g stated in the following theorem can be replaced by other conditions leading to the same restriction for H.

In the next theorem we use that

$$\{z \in \mathbb{R}^N \mid D(f, z) \neq \emptyset\} = D_0(f) + R(\hat{L}^{-1}\overline{L}),\qquad(6.1.22)$$

with $D_0(f)$ defined before Lemma 5.2.1. To see this, note that by definition of $D(f)$ in (6.1.5) the relation $\varepsilon \in D(f, z)$ holds if and only if $\overline{L}\varepsilon - \hat{L}z \in D(g)$, which, because of $D_0(f) = -\hat{L}^{-1}\big(D(g)\big)$, is equivalent to

$$z \in -\hat{L}^{-1}\big(D(g)\big) + \hat{L}^{-1}\overline{L}\varepsilon = D_0(f) + \hat{L}^{-1}\overline{L}\varepsilon.$$

Equation (6.1.22) is a consequence of these equivalences.

The following notation is used in the next theorem. For a vector field $g : \mathbb{R}^N \to \mathbb{R}^N$ and a linear mapping $B : \mathbb{R}^N \to S^3$ let

$$B\nabla g(z) : \mathbb{R}^N \to S^3$$

be the composition of B and the linear mapping defined by the $N \times N$-matrix $\nabla g(z)$, the Jacobi matrix at $z \in D(g)$. This composition is of course equal to $(Bg)'(z)$. The kernel of this map is a subspace of $\mathbb{R}^N$ with dimension greater or equal to $N - 6$.

Theorem 6.1.3 *Let $f, g, B, \overline{L}, \hat{L}$ be the functions and linear mappings from* (6.1.2) *– (6.1.4) with*

$$f(\varepsilon, z) = g(\overline{L}\varepsilon - \hat{L}z).$$

Suppose that the vector field g satisfies the following condition. For all $z \in \mathbb{R}^N$ with $D(f, z) \neq \emptyset$,

$$\bigcap_{\varepsilon \in D(f,z)} \ker(B\nabla g(\overline{L}\varepsilon - \hat{L}z)) = \{0\}.\qquad(6.1.23)$$

Then, if $(f, B) \in \mathcal{TM}^*$, for every vector field H which transforms this pair to pre-monotone type exists a bijective linear mapping $A : \mathbb{R}^N \to \mathbb{R}^N$ such that AH has the form (6.1.14) *on $D_0(f) + R(\hat{L}^{-1}\overline{L})$. Hence*

$$AH(z) = Pz + \tilde{H}(Qz)\qquad(6.1.24)$$

for all $z \in D_0(f) + R(\hat{L}^{-1}\overline{L})$. AH transforms the pair (f, B) to pre-monotone type. The transformed pair has the form

$$(f_{AH}, B(AH)^{-1}) = \big(g_{AH}(-\rho\nabla_h\psi_{AH}(\varepsilon, h)), B\big),\qquad(6.1.25)$$

where

$$-\rho\nabla_h\psi_{AH}(\varepsilon, h) = \overline{M}\varepsilon - Mh\qquad(6.1.26)$$

with $\overline{M} = B^T D$ and a symmetric, positive definite linear mapping M. These mappings satisfy

$$M^{-1}\overline{M} = \hat{L}^{-1}\overline{L}.\qquad(6.1.27)$$

If H transforms (f, B) to gradient type, then so does AH.

Proof: (I) Let H be a vector field which transforms (f, B) to pre-monotone type. Then the constitutive equations obtained by transformation of (6.1.2) – (6.1.4) with H are of the form (6.1.6) – (6.1.9), and H together with the symmetric, positive definite matrix M_H from (6.1.8) must satisfy conditions (i) – (iii) of Theorem 5.3.1. According to condition (i), the mapping BH^{-1} must be linear. We first collect the information about H that can be obtained from this condition.

Since $BH^{-1} : \mathbb{R}^N \to \mathcal{S}^3$ is linear, $\ker(BH^{-1})$ is a linear subspace of $\mathbb{R}^N$, and if we replace A in Lemma 6.1.1 by BH^{-1}, then this lemma shows that $\ker(BH^{-1})$ and the range

$$R(M_H^{-1}\overline{M}_H) = R(M_H^{-1}(BH^{-1})^T D)$$

are complementary subspaces of $\mathbb{R}^N$. Here $\overline{M}_H = (BH^{-1})^T D$ is the mapping from (6.1.9). Let $P_1 : \mathbb{R}^N \to \mathbb{R}^N$ be the projector onto $R(M_H^{-1}\overline{M}_H)$ along $\ker(BH^{-1})$, and let $Q_1 = I - P_1$. Then

$$P_1(\mathbb{R}^N) = R(M_H^{-1}\overline{M}_H), \quad \ker(P_1) = \ker(BH^{-1}) \tag{6.1.28}$$
$$Q_1(\mathbb{R}^N) = \ker(BH^{-1}), \quad \ker(Q_1) = R(M_H^{-1}\overline{M}_H). \tag{6.1.29}$$

From the linearity of $(BH)^{-1}$ we thus obtain for $z \in \mathbb{R}^N$

$$
\begin{aligned}
Bz &= BH^{-1}(H(z)) = BH^{-1}(P_1 + Q_1)H(z) \\
&= (BH^{-1})P_1 H(z) + (BH^{-1})Q_1 H(z) = (BH^{-1})P_1 H(z). \tag{6.1.30}
\end{aligned}
$$

We use now that $(BH^{-1})|_{R(M_H^{-1}\overline{M}_H)}$ is one-to-one, since $R(M_H^{-1}\overline{M}_H)$ is a complementary space of $\ker(BH^{-1})$. Therefore, because $P_1 H(z) \in R(M_H^{-1}\overline{M}_H)$, we conclude from (6.1.30) that $P_1 H$ is a linear mapping und that $(P_1 H)(z) = 0$ if and only if $Bz = 0$, hence

$$\ker(P_1 H) = \ker(B).$$

Since Q_1 maps into $\ker(BH^{-1})$ and since $P_1 + Q_1 = I$, it follows that

$$H(z) = (P_1 + Q_1)H(z) = P_1 H(z) + Q_1 H(z) \in \ker(BH^{-1})$$

if and only if $P_1 H(z) = 0$, and therefore if and only if $z \in \ker(P_1 H) = \ker(B)$. Because H is bijective, this means that

$$H(\ker(B)) = \ker(BH^{-1}).$$

The linearity of $P_1 H$ implies for all $z \in \mathbb{R}^N$ that $(P_1 H)'(z) = P_1 H$, which yields

$$
\begin{aligned}
H'(z) &= (P_1 + Q_1)H'(z) = P_1 H'(z) + Q_1 H'(z) \\
&= (P_1 H)'(z) + Q_1 H'(z) = P_1 H + Q_1 H'(z).
\end{aligned}
$$

For $w, z \in \mathbb{R}^N$ it thus follows as above that

$$[H'(z)]w = (P_1 H)w + [Q_1 H'(z)]w \in \ker(BH^{-1})$$

if and only if $w \in \ker(P_1 H) = \ker(B)$, which means

$$[H'(z)](\ker(B)) = \ker(BH^{-1}),$$

where we used that the linear map $H'(z)$ is bijective.

For later use we sum up the properties of H just proved. P_1H is linear with

$$\ker(P_1H) = \ker(B), \tag{6.1.31}$$

$$H(\ker(B)) = \ker(BH^{-1}), \tag{6.1.32}$$

$$[H'(z)](\ker(B)) = \ker(BH^{-1}), \quad \text{for all } z \in \mathbb{R}^N. \tag{6.1.33}$$

(II) To get more information about H, we now use condition (iii) of Theorem 5.3.1. H und M_H must satisfy this condition, and therefore also equation (5.3.5). If we insert the linear mapping

$$\overline{M}_H = (BH^{-1})^T D : \mathcal{S}^3 \to \mathbb{R}^N$$

from (6.1.9) into (5.3.5) and use that the mapping f from the constitutive equations (6.1.2) – (6.1.4) has the form

$$f(\varepsilon, z) = g(\overline{L}\varepsilon - \hat{L}z),$$

then (5.3.5) takes the form

$$H'(z)\nabla g(\overline{L}\varepsilon - \hat{L}z)[\overline{L} - \hat{L}H'(z)^{-1}M_H^{-1}\overline{M}_H]$$
$$+ H''(z) : g(\overline{L}\varepsilon - \hat{L}z)H'(z)^{-1}M_H^{-1}\overline{M}_H = 0 \tag{6.1.34}$$

for all $(\varepsilon, z) \in D(f)$. We start the investigation of this equation and first show that

$$P_1H''(z) : g(\overline{L}\varepsilon - \hat{L}z)H'(z)^{-1}M_H^{-1}\overline{M}_H = 0. \tag{6.1.35}$$

To this end let $P_1 = (p_{\ell i})_{\ell, i = 1, \ldots, N}$ be the $N \times N$-matrix representing the linear operator P_1, and let $g = (g_1, \ldots, g_N)$. Then the definition of the tensor H'' in Lemma 5.2.1 yields for $z \in \mathbb{R}^N, w \in D(g)$,

$$P_1H''(z) : g(w)$$

$$= \left(\sum_{i,k=1}^N p_{\ell i} \frac{\partial^2}{\partial z_k \partial z_j} H_i(z) g_k(w) \right)_{\ell, j = 1, \ldots, N}$$

$$= \left(\frac{\partial}{\partial z_j} \Big[\sum_{i,k=1}^N p_{\ell i} \frac{\partial}{\partial z_k} H_i(z) g_k(w) \Big] \right)_{\ell, j = 1, \ldots, N} \tag{6.1.36}$$

$$= \left(\frac{\partial}{\partial z_1} P_1 H'(z) g(w), \ldots, \frac{\partial}{\partial z_N} P_1 H'(z) g(w) \right).$$

Now, since P_1H is linear, it follows that

$$P_1H'(z)g(w) = [(P_1H)'(z)]g(w) = (P_1H)g(w)$$

is independent of z, which shows that the last term of (6.1.36) and therefore $P_1H''(z) : g(w)$ vanishes. This proves (6.1.35).

By multiplication of (6.1.34) with P_1 we obtain from (6.1.35) that for all $(\varepsilon, z) \in D(f)$ and $\eta \in \mathcal{S}^3$

$$0 = P_1H'(z)\nabla g(\overline{L}\varepsilon - \hat{L}z)[\overline{L} - \hat{L}H'(z)^{-1}M_H^{-1}\overline{M}_H]\eta$$

$$= (P_1H)\Big(\nabla g(\overline{L}\varepsilon - \hat{L}z)[\overline{L} - \hat{L}H'(z)^{-1}M_H^{-1}\overline{M}_H]\eta\Big).$$

Together with (6.1.31) this equation yields

$$(B\nabla g(\overline{L}\varepsilon - \hat{L}z))[\overline{L} - \hat{L}H'(z)^{-1}M_H^{-1}\overline{M}_H]\eta = 0.$$

Hence, using the assumption (6.1.23) for g,

$$[\overline{L} - \hat{L}H'(z)^{-1}M_H^{-1}\overline{M}_H]\eta \in \bigcap_{\varepsilon \in D(f,z)} \ker(B\nabla g(\overline{L}\varepsilon - \hat{L}z)) = \{0\}$$

for all z with $D(f,z) \neq \emptyset$ and all $\eta \in \mathcal{S}^3$. This implies $\overline{L} - \hat{L}H'(z)^{-1}M_H^{-1}\overline{M}_H = 0$, whence

$$H'(z)\hat{L}^{-1}\overline{L} = M_H^{-1}\overline{M}_H \tag{6.1.37}$$

for all $z \in D_0(f) + R(\hat{L}^{-1}\overline{L})$, where we noted (6.1.22).

(III) With this equation we can finish the investigation of H. (6.1.37) and (6.1.33) imply

$$H'(z)(R(\hat{L}^{-1}\overline{L})) = R(M_H^{-1}\overline{M}_H), \quad \text{for all } z \in D_0(f) + R(\hat{L}^{-1}\overline{L}) \tag{6.1.38}$$

$$H'(z)(\ker(B)) = \ker(BH^{-1}), \quad \text{for all } z \in \mathbb{R}^N.$$

We choose an arbitrary element $z \in D_0(f) + R(\hat{L}^{-1}\overline{L})$. For this z both of these equations hold, and since $R(M_H^{-1}\overline{M}_H)$ and $\ker(BH^{-1})$ are complementary subspaces, and since the linear mapping $H'(z)$ is invertible, they together imply that the range $R(\hat{L}^{-1}\overline{L})$ and $\ker(B)$ are also complementary subspaces of $\mathbb{R}^N$. Therefore the projector $P : \mathbb{R}^N \to \mathbb{R}^N$ onto $R(\hat{L}^{-1}\overline{L})$ along $\ker(B)$ can be defined. P and the projector $Q = I - P$ satisfy

$$P(\mathbb{R}^N) = R(\hat{L}^{-1}\overline{L}), \quad \ker(P) = \ker(B) \tag{6.1.39}$$

$$Q(\mathbb{R}^N) = \ker(B), \quad \ker(Q) = R(\hat{L}^{-1}\overline{L}). \tag{6.1.40}$$

From (6.1.39), (6.1.38), (6.1.28) and the linearity of P_1H we infer that for all $z \in D_0(f) + R(\hat{L}^{-1}\overline{L})$

$$H'(z)P = P_1H'(z)P = [(P_1H)'(z)]P = (P_1H)P.$$

We also note that (6.1.40) and (6.1.31) yield $(P_1H)Q = 0$. Hence

$$H'(z)P = (P_1H)P + (P_1H)Q = (P_1H)(P + Q) = P_1H. \tag{6.1.41}$$

Equation (6.1.39) implies that if $z \in D_0(f) + R(\hat{L}^{-1}\overline{L})$ and $t \in \mathbb{R}$ then also

$$tPz + Qz = (P + Q)z + (t - 1)Pz = z + (t - 1)Pz \in D_0(f) + R(\hat{L}^{-1}\overline{L}),$$

which means that if $z \in D_0(f) + R(\hat{L}^{-1}\overline{L})$, then (6.1.41) also holds with z replaced by $tPz + Qz$. This yields

$$\frac{d}{dt}H(tPz + Qz) = H'(tPz + Qz)Pz = (P_1H)z,$$

and consequently

$$H(z) = \int_0^1 \frac{d}{dt} H(tPz + Qz)dt + H(Qz) \tag{6.1.42}$$

$$= \int_0^1 (P_1 H)z \, dt + H(Qz) = (P_1 H)z + H(Qz),$$

for all $z \in D_0(f) + R(\hat{L}^{-1}\overline{L})$. If we observe that (6.1.40) and (6.1.31), (6.1.32) imply

$$P_1 H(Qz) = 0, \quad H(Qz) \in \ker(BH^{-1}), \tag{6.1.43}$$

and if we take (6.1.29) into account, we can write (6.1.42) equivalently as

$$H(z) = (P_1 H)(Pz + Qz) + H(Qz)$$
$$= P_1 H(Pz) + P_1 H(Qz) + Q_1 H(Qz) \tag{6.1.44}$$
$$= P_1 H(Pz) + Q_1 H(Qz).$$

The range $R(\hat{L}^{-1}\overline{L})$ is a complementary subspace of $\ker(B) = \ker(P_1 H)$, and therefore the linear mapping $P_1 H|_{R(\hat{L}^{-1}\overline{L})}$ is one-to-one. The range of this mapping is equal to $R(M_H^{-1}\overline{M}_H)$. This follows from

$$R(M_H^{-1}\overline{M}_H) = P_1(\mathbb{R}^N) = P_1(H(\mathbb{R}^N))$$
$$= (P_1 H)(P + Q)(\mathbb{R}^N) = (P_1 HP + P_1 HQ)(\mathbb{R}^N)$$
$$= P_1 H \, P(\mathbb{R}^N) = P_1 H|_{R(\hat{L}^{-1}\overline{L})} P(\mathbb{R}^N),$$

where we again used the first equation of (6.1.43) and (6.1.28), (6.1.39). Consequently,

$$P_1 H|_{R(\hat{L}^{-1}\overline{L})} : R(\hat{L}^{-1}\overline{L}) \to R(M_H^{-1}\overline{M}_H)$$

is a linear, bijective mapping. Moreover, from (6.1.33) it is seen that the subspaces $(\ker B)$ and $\ker(BH^{-1})$ are isomorphic. Therefore we can choose a bijective linear mapping $A : \mathbb{R}^N \to \mathbb{R}^N$ with

$$A(R(M_H^{-1}\overline{M}_H)) = R(\hat{L}^{-1}\overline{L})$$
$$A(\ker(BH^{-1})) = \ker(B),$$

which acts as the inverse of $P_1 H|_{R(\hat{L}^{-1}\overline{L})}$ on $R(M_H^{-1}\overline{M}_H)$. From (6.1.28), (6.1.29) and (6.1.40) we then have $AQ_1 = QA$, and thus obtain from (6.1.44) for $z \in D_0(f) + R(\hat{L}^{-1}\overline{L})$ that

$$AH(z) = AP_1 H(Pz) + AQ_1 H(Qz)$$
$$= Pz + QAH(Qz) = Pz + \tilde{H}(Qz),$$

where we define $\tilde{H} : \ker(B) \to \ker(B) \subseteq \mathbb{R}^N$ by

$$\tilde{H}(z) = QAH(z) = AH(z).$$

Thus we have demonstrated that A can be found such that AH is of the form (6.1.24).

That the pair (f_{AH}, B) obtained by transformation of (f, B) with AH is of pre-monotone type is an obvious consequence of Lemma 5.2.2, since one immediately convinces oneself that by definition of the transformation of pairs in and after Lemma 5.1.1,

$$(f_{AH}, B) = (f_{AH}, B(AH)^{-1}) = \left((f_H)_A, (BH^{-1})A^{-1}\right).$$

This means that (f_{AH}, B) is obtained by transformation of (f_H, BH^{-1}), which by assumption belongs to $\mathcal{M}^*$, with the linear vector field A. Lemma 5.2.2 thus implies that also $(f_{AH}, B) \in \mathcal{M}^*$, and also that $(f_{AH}, B) \in \mathcal{G}$, if $(f_H, BH^{-1}) \in \mathcal{G}$.

To establish (6.1.27) we note that the same considerations as for H can be carried through for AH. Instead of (6.1.37) we then obtain

$$(AH)'(z)\hat{L}^{-1}\overline{L} = M^{-1}\overline{M}. \tag{6.1.45}$$

Using that (6.1.39), (6.1.40) imply $P\hat{L}^{-1}\overline{L} = \hat{L}^{-1}\overline{L}$ and $Q\hat{L}^{-1}\overline{L} = 0$, and that (6.1.24) yields

$$(AH)'(z) = P + \tilde{H}'(Qz)Q,$$

we obtain from (6.1.45)

$$M^{-1}\overline{M} = P\hat{L}^{-1}\overline{L} + \tilde{H}'(Qz)Q\hat{L}^{-1}\overline{L} = \hat{L}^{-1}\overline{L},$$

which proves (6.1.27). The proof of Theorem 6.1.3 is complete.

If we combine the last two theorems, we obtain the following

Corollary 6.1.4 *Assume that the vector field* $g : \mathbb{R}^N \to \mathbb{R}^N$ *from equation* (6.1.4) *satisfies*

$$\bigcap_{\varepsilon \in D(f,z)} \ker(B\nabla g(\overline{L}\varepsilon - \hat{L}z)) = \{0\} \tag{6.1.46}$$

for all $z \in \mathbb{R}^N$ *with* $D(f, z) \neq \emptyset$. *Then:*

(I) *The following statements* (1), (2), (3) *are equivalent:*
  (1) *There exists a symmetric, positive definite* $N \times N$*-matrix* M *satisfying*

  (i) $\quad M^{-1}\overline{M} = \hat{L}^{-1}\overline{L}$, $\quad$ *where* $\overline{M} = B^T D$,
  (ii) $\quad L = M - B^T DB$ $\quad$ *is positive semi-definite.*

  (2) *The constitutive equations* (6.1.2) - (6.1.4) *are of pre-monotone type.*
  (3) *The constitutive equations* (6.1.2) - (6.1.4) *can be transformed to pre-monotone type.*

(II) *A vector field* $H : \mathbb{R}^N \to \mathbb{R}^N$ *transforms these constitutive equations to pre-monotone type if for all* $z \in \mathbb{R}^N$ *it is of the form*

$$H(z) = A^{-1}(P(z) + \tilde{H}(Qz)), \tag{6.1.47}$$

where $A : \mathbb{R}^N \to \mathbb{R}^N$ *is a bijective linear mapping, where* P *is the projector along* $\ker(B)$ *onto the range*

$$R(\hat{L}^{-1}\overline{L}) = \{\hat{L}^{-1}\overline{L}\varepsilon \in \mathbb{R}^N \mid \varepsilon \in \mathcal{S}^3\},$$

where $Q = I - P$, and where $\tilde{H} : \ker(B) \to \ker(B) \subseteq \mathbb{R}^N$ is bijective as a mapping of $\ker(B)$ into $\ker(B)$. Conversely, a vector field $H : \mathbb{R}^N \to \mathbb{R}^N$ must be for all $z \in D_0(f) + R(\hat{L}^{-1}\overline{L})$ of the form (6.1.47), if it transforms these constitutive equations to pre-monotone type.

Proof. Theorem 6.1.2 shows that if conditions (i) and (ii) are satisfied, then (6.1.2) - (6.1.4) are of pre-monotone type, and thus, in particular, can be transformed to pre-monotone type. Theorem 6.1.3 shows that if (6.1.2) - (6.1.4) can be transformed to pre-monotone type, then there exists a symmetric, positive definite matrix M which satisfies condition (i) of the corollary. It is defined by $Mh = \rho\nabla_h\psi(0, h)$ for a free energy ψ from pre-monotone-type equations, and thus, by Definition 3.1.1, satisfies condition (ii).

Moreover, Theorem 6.1.2 shows that if (i) and (ii) are satisfied, then all vector fields of the form $Pz + \tilde{H}(Qz)$ transform (6.1.2) - (6.1.4) to pre-monotone type, and therefore, as a consequence of Lemma 5.2.2, all vector fields of the form (6.1.47). Conversely, Theorem 6.1.3 shows that if the vector field H transforms (6.1.2) - (6.1.4) to pre-monotone type then it must be of the form (6.1.47) on $D_0(f) + R(\hat{L}^{-1}\overline{L})$.

6.2 Transformation of Pre-Monotone Equations to Gradient Type

We study now the transformation of the constitutive equations (6.1.2) - (6.1.4) to gradient type. Since $\mathcal{G} \subseteq \mathcal{M}^*$, a vector field which transforms these equations to gradient type transforms them in particular to pre-monotone type. By Corollary 6.1.4 it must have the form (6.1.47) on $D_0(f) + R(\hat{L}^{-1}\overline{L})$, if condition (6.1.46) is satisfied for the given constitutive equations. By composition with a suitable bijective linear map, we thus obtain a vector field of the form

$$H(z) = Pz + \tilde{H}(Qz) \tag{6.2.1}$$

for $z \in D_0(f) + R(\hat{L}^{-1}\overline{L})$. Lemma 5.2.2 shows that equations (6.1.2) - (6.1.4) are transformed to gradient type by the original vector field if and only if they are transformed to gradient type by the vector field (6.2.1). Therefore, since we want to derive conditions which determine the subclass of those equations (6.1.2) - (6.1.4) that can be transformed to gradient type, it suffices to study when these equations can be transformed to gradient type by vector fields of the form (6.2.1).

Thus, let $H : \mathbb{R}^N \to \mathbb{R}^N$ be a vector field which transforms (6.1.2) - (6.1.4) to gradient type, and which satisfies (6.2.1) on $D_0(f) + R(\hat{L}^{-1}\overline{L})$ with P, Q and $\tilde{H} : \ker(B) \to \ker(B) \subseteq \mathbb{R}^N$ defined as in Corollary 6.1.4. Comparison with Theorem 6.1.3 shows that the transformed equations are of the form

$$T = D(\varepsilon - Bh)$$
$$h_t = \nabla\chi(-\rho\nabla_h\psi(\varepsilon, h)),$$

where the quadratic form ψ satisfies

$$-\rho\nabla_h\psi(\varepsilon, h) = \overline{M}\varepsilon - Mh = B^T D\varepsilon - Mh,$$

with M satisfying conditions (i) and (ii) of Corollary 6.1.4. Theorem 5.4.1 yields that H and the function $\varphi = \chi \circ (-M) \circ H$ must satisfy the system of partial differential equations

$$- H'(z)^T M H'(z) f(0, z) = \nabla \varphi(z) \qquad (6.2.2)$$

for all $z \in D_0(f)$. Since $R(\hat{L}^{-1}\overline{L})$ is a subspace of $\mathbb{R}^N$, it follows that $D_0(f) \subseteq D_0(f) + R(\hat{L}^{-1}\overline{L})$, whence H satisfies (6.2.1) for all $z \in D_0(f)$. We infer that transformation of (6.1.2) - (6.1.4) to gradient type is only possible if the system (6.2.2) has solutions of the form (6.2.1). One cannot expect that this is always the case, and in this section we investigate under what conditions vector fields of such form can be solutions of this system of partial differential equations.

In the following we always assume that $R(\hat{L}^{-1}\overline{L})$ and $\ker(B)$ are complementary subspaces. P and Q denote the projections defined in Theorem 6.1.2. To state the next lemma, we need some new definitions and notations. Let $P^\perp : \mathbb{R}^N \to \mathbb{R}^N$ and $Q^\perp : \mathbb{R}^N \to \mathbb{R}^N$ denote the orthogonal projections onto the subspaces $R(\hat{L}^{-1}\overline{L})$ and $\ker(B)$, respectively, and for a linear, symmetric, positive definite mapping $M : \mathbb{R}^N \to \mathbb{R}^N$ let the mappings $\hat{M}$, $\tilde{M}$ be defined by

$$\hat{M} = P^\perp M \big|_{R(\hat{L}^{-1}\overline{L})} : R(\hat{L}^{-1}\overline{L}) \to R(\hat{L}^{-1}\overline{L})$$

$$\tilde{M} = Q^\perp M \big|_{\ker(B)} : \ker(B) \to \ker(B).$$

For a continuously differentiable function $\varphi : \mathbb{R}^N \to \mathbb{R}$ we denote by $\hat{\nabla}\varphi$ and $\tilde{\nabla}\varphi$ the components of the gradient $\nabla \varphi$ lying in the subspaces $R(\hat{L}^{-1}\overline{L})$ and $\ker(B)$, respectively, hence

$$\hat{\nabla}\varphi(z) = P^\perp(\nabla\varphi(z)), \quad \tilde{\nabla}\varphi(z) = Q^\perp(\nabla\varphi(z)). \qquad (6.2.3)$$

To a linear mapping $X : \mathbb{R}^N \to \ker(B)$ let $X^* : \ker(B) \to \mathbb{R}^N$ be the dual mapping, that assigns to $w \in \ker(B)$ the unique element $X^* w \in \mathbb{R}^N$ with

$$(X^* w) \cdot z = w \cdot X z$$

for all $z \in \mathbb{R}^N$, where on the right hand side the scalar product on $\ker(B)$ is the restriction of the scalar product of $\mathbb{R}^N$ to the subspace $\ker(B)$. It is obvious that X^* coincides with the restriction to $\ker(B)$ of the transpose X^T of X, regarded however as a mapping $X : \mathbb{R}^N \to \mathbb{R}^N$. We can thus use the same notation and write X^T instead of X^*. In the following proof this convention is applied to the mapping Q^T and also to the mapping $(Q^\perp)^T = Q^\perp$, which as a mapping $Q^\perp : \ker(B) \to \mathbb{R}^N$ is the embedding of $\ker(B)$ into $\mathbb{R}^N$.

By definition, $\tilde{H}$ takes values in R^N, but has range $R(\tilde{H}) = \ker(B)$. Therefore we can regard $\tilde{H}$ as map $\hat{H} : \ker(B) \to \ker(B)$, and in the following lemma $\tilde{H}$ is understood in this sense. Consequently, for every $z \in \ker(B)$ the derivative of $\tilde{H}$ is a linear map $\tilde{H}'(z) : \ker(B) \to \ker(B)$, and in particular, $\tilde{H}'(z)^T$ means the transpose of $\tilde{H}'(z)$ on $\ker(B)$.

Lemma 6.2.1 *Let the linear, symmetric, positive definite mapping $M : \mathbb{R}^N \to \mathbb{R}^N$ in (6.2.2) satisfy $R(M^{-1}\overline{M}) = R(\hat{L}^{-1}\overline{L})$ with $\overline{M} = B^T D$. Then a transformation field H, which on $D_0(f) + R(\hat{L}^{-1}\overline{L})$ is of the form (6.2.1), and a continuously differentiable function $\varphi : \mathbb{R}^N \to \mathbb{R}$ solve (6.2.2) if and only if*

$$- \hat{M} P f(0, z) = \hat{\nabla}\varphi(z) \tag{6.2.4}$$

$$- Q^\perp \tilde{H}'(Qz)^T \tilde{M} \tilde{H}'(Qz) Q f(0, z) = \tilde{\nabla}\varphi(z) \tag{6.2.5}$$

for all $z \in D_0(f)$.

Proof. We first note that the projectors P, $Q^\perp$ satisfy

$$Q^\perp M P = 0, \quad P^T M Q^\perp = 0. \tag{6.2.6}$$

To obtain these equations observe that by assumption

$$
\begin{aligned}
MP(\mathbb{R}^N) &= M R(\hat{L}^{-1}\overline{L}) = M R(M^{-1}\overline{M}) = R(\overline{M}) = R(B^T D) \\
&= R((DB)^T) = [\ker(DB)]^\perp = \ker(B)^\perp,
\end{aligned}
$$

where in the last step we used that $D : \mathcal{S}^3 \to \mathcal{S}^3$ is positive definite and therefore invertible. (6.2.6) follows from this equation, since $\ker(Q^\perp) = \ker(B)^\perp$ implies $Q^\perp M P = 0$, whence $P^T M Q^\perp = (Q^\perp M P)^T = 0$.

After this preparation we can verify (6.2.4) and (6.2.5). To this end we write (6.2.1) in the form

$$H(z) = Pz + Q^\perp \tilde{H}(Qz),$$

which results in

$$H'(z) = P + Q^\perp \tilde{H}'(Qz) Q$$

and

$$H'(z)^T = P^T + Q^T \tilde{H}'(Qz)^T Q^\perp.$$

Together with (6.2.6) and with the definition of $\tilde{M}$ we thus obtain

$$
\begin{aligned}
H'(z)^T M H'(z) \\
= \left[P^T + Q^T \tilde{H}'(Qz)^T Q^\perp \right] M \left[P + Q^\perp \tilde{H}'(Qz) Q \right] \\
= P^T M P + P^T M Q^\perp \tilde{H}'(Qz) Q \\
+ Q^T \tilde{H}'(Qz)^T \left[Q^\perp M P + Q^\perp M Q^\perp \tilde{H}'(Qz) Q \right] \\
= P^T M P + Q^T \tilde{H}'(Qz)^T \tilde{M} \tilde{H}'(Qz) Q.
\end{aligned}
\tag{6.2.7}
$$

We now use

$$P^\perp Q^T = Q^\perp P^T = 0, \tag{6.2.8}$$

which follows from $R(Q^T) = \ker(Q)^\perp = R(\hat{L}^{-1}\overline{L})^\perp = \ker(P^\perp)$ and from $R(P^T) = \ker(P)^\perp = \ker(B)^\perp = \ker(Q^\perp)$. Multiplication of (6.2.7) on the left by $P^\perp$ and observation of (6.2.8) results in

$$P^\perp H'(z)^T M H'(z) = P^\perp P^T M P = P^\perp (P^T + Q^T) M P = P^\perp M P = \hat{M} P,$$

whereas multiplication by $Q^\perp$ and application of (6.2.8) leads to

$$Q^\perp H'(z)^T M H'(z)$$

$$= Q^\perp Q^T \tilde{H}'(Qz)^T \tilde{M} \tilde{H}'(Qz)Q$$

$$= Q^\perp \big(Q^T + P^T\big) \tilde{H}'(Qz)^T \tilde{M} \tilde{H}'(Qz)Q$$

$$= Q^\perp \tilde{H}'(Qz)^T \tilde{M} \tilde{H}'(Qz)Q.$$

Multiplication of (6.2.2) on the left by $P^\perp$ or $Q^\perp$ and insertion of the above expressions yields the statement of the lemma. The proof is complete.

To simplify the notation, we transform the equations (6.2.4), (6.2.5) to new coordinates in $\mathbb{R}^N$. To introduce these coordinates, let d be the dimension of the subspace $R(M^{-1}\overline{M}) = R(\hat{L}^{-1}\overline{L})$, hence $\dim(\ker(B)) = N - d$. From $\overline{M} = B^T D$ it follows that

$$d = \dim R(B^T) = \dim R(B) \leq 6.$$

The new coordinates are denoted by

$$(\hat{z}, \tilde{z}) = (\hat{z}_1, \ldots, \hat{z}_d, \tilde{z}_1, \ldots, \tilde{z}_{N-d}) \in \mathbb{R}^N.$$

They are chosen such that the coordinate transformation

$$z \mapsto Jz = (\hat{z}, \tilde{z}) : \mathbb{R}^N \to \mathbb{R}^N$$

from the old to the new coordinates is linear, such that

$$R(\hat{L}^{-1}\overline{L}) = \{(\hat{z}, 0) \mid \hat{z} \in \mathbb{R}^d\}, \quad \ker(B) = \{(0, \tilde{z}) \mid \tilde{z} \in \mathbb{R}^{N-d}\},$$

and such that the mappings

$$z \mapsto \hat{J}z = \hat{z} : R(\hat{L}^{-1}\overline{L}) \to \mathbb{R}^d$$

$$z \mapsto \tilde{J}z = \tilde{z} : \ker(B) \to \mathbb{R}^{N-d}$$

are unitary, hence $\hat{J}^T = \hat{J}^{-1}$ and $\tilde{J}^T = \tilde{J}^{-1}$. With the mappings $\hat{J}$ and $\tilde{J}$ we can write J in the form

$$Jz = (\hat{J}Pz, \tilde{J}Qz). \tag{6.2.9}$$

In the following proof it is convenient to write this equation in operator form. To this end let $\hat{P} : \mathbb{R}^N \to \mathbb{R}^N$ and $\tilde{Q} : \mathbb{R}^N \to \mathbb{R}^N$ denote the orthogonal projections onto the subspaces $\mathbb{R}^d \times \{0\}$ and $\{0\} \times \mathbb{R}^{N-d}$, respectively. The mappings $\hat{P}$ and $\tilde{Q}$ are equal to their transposes, and as above, the duals of $\hat{P}$ and $\tilde{Q}$, when regarded as mappings of $\mathbb{R}^N$ onto $\mathbb{R}^d \times \{0\}$ and $\{0\} \times \mathbb{R}^{N-d}$, respectively, coincide with the restrictions of $\hat{P}$ and $\tilde{Q}$ to these subspaces, and are therefore also denoted by $\hat{P}$ and $\tilde{Q}$. With these projections we can write (6.2.9) in the form

$$J = \hat{P}\hat{J}P + \tilde{Q}\tilde{J}Q. \tag{6.2.10}$$

Using this equation one immediately verifies that the inverse and the transpose of the coordinate transformation J satisfy

$$J^{-1} = P\hat{J}^{-1}\hat{P} + Q\tilde{J}^{-1}\tilde{Q}$$

$$J^T = P^T \hat{J}^{-1}\hat{P} + Q^T \tilde{J}^{-1}\tilde{Q}.$$

In the new coordinates the function f is denoted by

$$f(\varepsilon, \hat{z}, \tilde{z}) = (\hat{f}(\varepsilon, \hat{z}, \tilde{z}), \tilde{f}(\varepsilon, \hat{z}, \tilde{z}))$$

with

$$\hat{f}(\varepsilon, \hat{z}, \tilde{z}) = \hat{J}Pf(\varepsilon, J^{-1}(\hat{z}, \tilde{z})) \in \mathbb{R}^d$$

$$\tilde{f}(\varepsilon, \hat{z}, \tilde{z}) = \tilde{J}Qf(\varepsilon, J^{-1}(\hat{z}, \tilde{z})) \in \mathbb{R}^{N-d}.$$

As in the preceding lemma, we regard $\tilde{H}$ as function $\tilde{H} : \ker(B) \to \ker(B)$. The mappings obtained by transformation of this function and of $\hat{M}$, $\tilde{M}$ and φ to new coordinates are again denoted by $\tilde{H}$, $\hat{M}$, $\tilde{M}$ and φ. In particular, this means that

$$\tilde{z} \mapsto \tilde{H}(\tilde{z}) = \tilde{J}H(\tilde{J}^{-1}\tilde{z}) : \mathbb{R}^{N-d} \to \mathbb{R}^{N-d}, \quad (\hat{z}, \tilde{z}) \mapsto \varphi(\hat{z}, \tilde{z}) = \varphi(J^{-1}(\hat{z}, \tilde{z})),$$

$$\hat{M} : \mathbb{R}^d \to \mathbb{R}^d, \quad \tilde{M} : \mathbb{R}^{N-d} \to \mathbb{R}^{N-d}.$$

Now we can state the differential equations from the last lemma in the new coordinates:

Lemma 6.2.2 *In the new coordinates, the equations (6.2.4) and (6.2.5) are given by*

$$-\hat{M}\hat{f}(0, \hat{z}, \tilde{z}) = \nabla_{\hat{z}}\varphi(\hat{z}, \tilde{z}) \tag{6.2.11}$$

$$-\tilde{H}'(\tilde{z})^T \tilde{M} \tilde{H}'(\tilde{z}) \, \tilde{f}(0, \hat{z}, \tilde{z}) = \nabla_{\tilde{z}}\varphi(\hat{z}, \tilde{z}), \tag{6.2.12}$$

for all $(\hat{z}, \tilde{z}) \in D_0(f)$.

Proof. Contrary to the above introduced convention, in this proof we denote the mappings obtained by transformation of $\tilde{H}$, $\hat{M}$, $\tilde{M}$ and φ to new coordinates by $\tilde{H}_J$, $\hat{M}_J$, $\tilde{M}_J$ and φ_J. Since we regard gradients of real valued functions as column vectors, the equation $\varphi(z) = \varphi_J(Jz)$ yields

$$J\tilde{\nabla}\varphi(z) = JQ^\perp \nabla\varphi(z) = JQ^\perp \left[\nabla\varphi_J(Jz)^T J\right]^T = JQ^\perp J^T \nabla\varphi_J(Jz). \tag{6.2.13}$$

The definition of J in (6.2.10) yields

$$JQ^\perp J^T = \left[\hat{P}\hat{J}P + \tilde{Q}\tilde{J}Q\right]Q^\perp\left[P^T\hat{J}^{-1}\hat{P} + Q^T\tilde{J}^{-1}\tilde{Q}\right]$$

$$= \tilde{Q}\tilde{J}QQ^\perp Q^T \tilde{J}^{-1}\tilde{Q} = \tilde{Q}\tilde{J}Q^\perp Q^T \tilde{J}^{-1}\tilde{Q} \tag{6.2.14}$$

$$= \tilde{Q}\tilde{J}Q^\perp\left(Q^T + P^T\right)\tilde{J}^{-1}\tilde{Q} = \tilde{Q}\tilde{J}Q^\perp\tilde{J}^{-1}\tilde{Q} = \tilde{Q}\tilde{J}\tilde{J}^{-1}\tilde{Q} = \tilde{Q},$$

where we used that $Q^\perp P^T = PQ^\perp = 0$, which follows from $Q^\perp(\mathbb{R}^N) = \ker(B) = \ker(P)$ and from $R(P^T) = \ker(P)^\perp = \ker(Q^\perp)$. Equations (6.2.13) and (6.2.14) together yield

$$J\tilde{\nabla}\varphi(z) = \tilde{Q}\nabla\varphi_J(Jz) = (0, \nabla_{\tilde{z}}\varphi_J(Jz)). \tag{6.2.15}$$

In exactly the same way we obtain

$$J\hat{\nabla}\varphi(z) = (\nabla_{\hat{z}}\varphi_J(Jz), 0). \tag{6.2.16}$$

In the following we denote the Jacobi-matrix of $\tilde{H}_J$ by $\tilde{H}'_J$. From the equation $\tilde{H}(z) = \tilde{J}^{-1}\tilde{H}_J(\tilde{J}z)$, which holds for $z \in \ker(B)$, we infer that

$$\tilde{H}'(Qz)^T = \left[\tilde{J}^{-1}\tilde{H}'_J(\tilde{J}Qz)\tilde{J}\right]^T = \tilde{J}^{-1}\tilde{H}'_J(\tilde{J}Qz)^T\tilde{J},$$

hence

$$JQ^\perp \tilde{H}'(Qz)^T \tilde{M} \tilde{H}'(Qz)Qf(0,z)$$
$$= JQ^\perp \tilde{J}^{-1}\tilde{H}'_J(\tilde{J}Qz)^T \tilde{J}\tilde{J}^{-1}\tilde{M}_J\tilde{J}\tilde{J}^{-1}\tilde{H}'_J(\tilde{J}Qz)\tilde{J}Qf(0,z) \qquad (6.2.17)$$
$$= JQ^\perp \tilde{J}^{-1}\tilde{H}'_J(\tilde{J}Qz)^T \tilde{M}_J \tilde{H}'_J(\tilde{J}Qz)\tilde{f}(0,Jz),$$

where we used that $\tilde{J}Qf(0,z) = \tilde{f}(0,Jz)$. Moreover,

$$JQ^\perp \tilde{J}^{-1} = [\hat{P}\tilde{J}P + \tilde{Q}\tilde{J}Q]Q^\perp \tilde{J}^{-1} = \tilde{Q}\tilde{J}QQ^\perp \tilde{J}^{-1} = \tilde{Q}\tilde{J}\tilde{J}^{-1} = \tilde{Q}, \qquad (6.2.18)$$

where $Q^\perp : \ker(B) \to \mathbb{R}^N$, $\tilde{Q} : \{0\}\times\mathbb{R}^{N-d} \to \mathbb{R}^N$ are the embeddings. Combination of (6.2.15), (6.2.17) and (6.2.18) with (6.2.5) results in

$$-(0, \tilde{H}'_J(\tilde{J}Qz)^T \tilde{M}_J \tilde{H}'_J(\tilde{J}Qz)\tilde{f}(0,Jz)) = (0, \nabla_{\tilde{z}}\varphi_J(Jz)).$$

Inserting $Jz = (\hat{z}, \tilde{z})$, $\tilde{J}Qz = \tilde{z}$, and taking the last $N - d$ components of this equation yields (6.2.12). Equation (6.2.11) is obtained by an analogous, but simpler calculation using (6.2.16). The proof of Lemma 6.2.2 is complete.

We are now in a position to state in the next theorem the classification conditions for the class $\mathcal{TG}$. We give these conditions under a hypotheses for the geometry of the set

$$D_0(f) = \{z \in \mathbb{R}^N \mid (0,z) \in D(f)\} = \{z \in \mathbb{R}^N \mid -\hat{L}z \in D(g)\},$$

where, as always in this chapter, f and g are the functions from the constitutive equations (6.1.3) and (6.1.4). To formulate the theorem, we need some definitions and notations:

Definition 6.2.3 (i) *A set $X \subseteq \mathbb{R}^m$ is called star-shaped with center 0, if to every $w \in X$ the segment $\{tw \mid 0 \leq t \leq 1\}$ belongs to X.*
(ii) *Let $\Gamma : \mathbb{R}^d \to \mathbb{R}^d$ be a bijective, continuously differentiable map, whose Jacobi matrix $\Gamma'(w)$ is non-singular for all $w \in \mathbb{R}^d$, and let $\hat{z}_0 = \Gamma(0)$. We say that a set $X \subseteq \mathbb{R}^d$ is Γ-star-shaped with center $\hat{z}_0$, if the set $\Gamma^{-1}(X)$ is star-shaped with center 0.*

We use the notations

$$\hat{D}_0(f,\tilde{z}) = \{\hat{z} \in \mathbb{R}^d \mid (\hat{z},\tilde{z}) \in D_0(f)\}, \quad \tilde{z} \in \mathbb{R}^{N-d}$$
$$\tilde{D}_0(f,\hat{z}) = \{\tilde{z} \in \mathbb{R}^{N-d} \mid (\hat{z},\tilde{z}) \in D_0(f)\}, \quad \hat{z} \in \mathbb{R}^d.$$

Finally, we define a vector field ω, whose domain of definition contains every section $\hat{D}_0(f,\tilde{z})$: Assume that there exists a mapping $\Gamma : \mathbb{R}^d \to \mathbb{R}^d$ such that for every

$\tilde{z} \in \mathbb{R}^{N-d}$ the sets $\hat{D}_0(f, \tilde{z})$ are Γ–star-shaped with center $\hat{z}_0 = \Gamma(0)$. Then, by definition, $\hat{z}_0 \in \hat{D}_0(f, \tilde{z})$ and for every $\hat{z} \in \hat{D}_0(f, \tilde{z}) \backslash \{\hat{z}_0\}$ with $w = \Gamma^{-1}(\hat{z})$ the path

$$t \mapsto \Gamma(tw) : [0, 1] \to \hat{D}_0(f, \tilde{z})$$

connects $\hat{z}_0$ and $\hat{z}$ within $\hat{D}_0(f, \tilde{z})$. Since

$$\frac{d}{dt} \Gamma(tw)\big|_{t=1} = \Gamma'(w)w \, ,$$

the vector field $\omega : R^d \backslash \{\hat{z}_0\} \to \mathbb{R}^d$ defined by

$$\omega(\hat{z}) = \omega(\Gamma(w)) = \Gamma'(w)w / |\Gamma'(w)w| \tag{6.2.19}$$

is tangential to every such path and therefore has as integral curves the paths $t \mapsto \Gamma(tw)$, which all start in $\hat{z}_0$ and cover every section $\hat{D}_0(f, \tilde{z})$.

We can now state the theorem:

Theorem 6.2.4 *Let*

$$T = D(\varepsilon - Bz)$$
$$z_t = f(\varepsilon, z) = g(\overline{L}\varepsilon - \hat{L}z)$$

be constitutive equations and assume that the coordinates are chosen as above such that

$$R(\hat{L}^{-1}\overline{L}) = \{(\hat{z}, 0) \mid \hat{z} \in \mathbb{R}^d\}, \quad \ker(B) = \{(0, \tilde{z}) \mid \tilde{z} \in \mathbb{R}^{N-d}\} \, .$$

Suppose that there exists a mapping $\Gamma : \mathbb{R}^d \to \mathbb{R}^d$ such that with $\hat{z}_0 = \Gamma(0)$ the following two assumptions are satisfied:

(i)   *The section $\tilde{D}_0(f, \hat{z}_0)$ is simply connected*

(ii) *For every $\tilde{z} \in \mathbb{R}^{N-d}$ the sections $\hat{D}_0(f, \tilde{z})$ are Γ–star-shaped with center $\hat{z}_0$.*

Then the following assertions (I) and (II) hold:

(I) *Let $M : \mathbb{R}^N \to \mathbb{R}^N$ be a linear, symmetric, positive definite mapping, which together with $\overline{M} = B^T D$ satisfies $R(M^{-1}\overline{M}) = R(\hat{L}^{-1}\overline{L})$. Then for a transformation field $H : \mathbb{R}^N \to \mathbb{R}^N$, which on $D_0(f) + R(\hat{L}^{-1}\overline{L})$ is of the form $H(z) = Pz + \tilde{H}(Qz)$, there exists a function $\varphi : \mathbb{R}^N \to \mathbb{R}$, which together with H satisfies*

$$- H'(z)^T M H'(z) \, f(0, z) = \nabla \varphi(z) \tag{6.2.20}$$

on $D_0(f)$, if and only if the following conditions (iii) and (iv) hold:

(iii) *$\hat{f}$ satisfies for all $1 \leq i, j \leq d$ and $z = (\hat{z}, \tilde{z}) \in D_0(f)$ the compatibility conditions*

$$\left[\hat{M} \frac{\partial}{\partial \hat{z}_i} \hat{f}(0, \hat{z}, \tilde{z})\right]_j = \left[\hat{M} \frac{\partial}{\partial \hat{z}_j} \hat{f}(0, \hat{z}, \tilde{z})\right]_i , \tag{6.2.21}$$

where $[\hat{M}(\partial \hat{f}/\partial \hat{z}_i)]_j$ denotes the j–th component of the vector $\hat{M}(\partial \hat{f}/\partial \hat{z}_i) \in \mathbb{R}^d$.

(iv)   *The vector field $\tilde{H}$ satisfies the two systems of partial differential equations*

$$\tilde{H}'(\tilde{z})^T \tilde{M} \tilde{H}'(\tilde{z}) \left(\omega(\hat{z}) \cdot \nabla_{\tilde{z}} \right) \hat{f}(0,\hat{z},\tilde{z}) = \nabla_{\tilde{z}} \left(\omega(\hat{z}) \cdot \hat{M} \hat{f}(0,\hat{z},\tilde{z}) \right) \qquad (6.2.22)$$

for all $z = (\hat{z}, \tilde{z}) \in D_0(f)$, *and*

$$\frac{\partial}{\partial \tilde{z}_i} \left[\tilde{H}'(\tilde{z})^T \tilde{M} \tilde{H}'(\tilde{z}) \hat{f}(0,\hat{z}_0,\tilde{z}) \right]_j = \frac{\partial}{\partial \tilde{z}_j} \left[\tilde{H}'(\tilde{z})^T \tilde{M} \tilde{H}'(\tilde{z}) \hat{f}(0,\hat{z}_0,\tilde{z}) \right]_i \quad (6.2.23)$$

for all $\tilde{z} \in \tilde{D}(f,\hat{z}_0)$ *and* $1 \leq i, j \leq N - d$.

(II) *The pair* (f, B) *belongs to the class* $\mathcal{TG}$, *if a symmetric, positive definite* $N \times N-$
matrix M *and a vector field* $\tilde{H} : \mathbb{R}^{N-d} \to \mathbb{R}^{N-d}$ *exist satisfying* (iii), (iv) *and the conditions:*

(v) $M^{-1}\overline{M} = \hat{L}^{-1}\overline{L}$, *where* $\overline{M} = B^T D$

(vi) $L = M - B^T D B$ *is positive semi-definite.*

(III) *Assume that in addition to* (i) *and* (ii) *the following assumption is satisfied:*

(vii)   *For all* $z \in \mathbb{R}^N$ *with* $D(f,z) = \{ \varepsilon \in \mathcal{S}^3 \mid (\varepsilon, z) \in D(f) \} \neq \emptyset$ *the relation*

$$\bigcap_{\varepsilon \in D(f,z)} \ker(B \nabla g(\overline{L}\varepsilon - \hat{L}z)) = \{0\}, \qquad (6.2.24)$$

holds, where g *is the vector field from the constitutive equations.*

Then the pair (f, B) *belongs to the class* $\mathcal{TG}$ *if and only if* M *and* $\tilde{H}$ *exist satisfying*
(iii) - (vi). *Under the assumptions* (i), (ii), (vii) *for* f *and* g, *the conditions* (iii) -
(vi) *are therefore classification conditions for the class* $\mathcal{TG}$.

Remark. Conditions (i), (ii) and (vii) for f and g are not the only ones, which
guarantee that (iii) - (vi) classify the class $\mathcal{TG}$, because (vii) can be replaced by other
conditions. Since (vii) suffices for the applications of this theorem in Chapter 8, we
shall not discuss these conditions here.

Proof. (I) By Lemmas 6.2.1 and 6.2.2, it suffices to show that the set of equations
(6.2.21) − (6.2.23) is equivalent to the system (6.2.11), (6.2.12). For brevity, in this
proof we use the notation

$$\mathcal{H}(\tilde{z}) = \tilde{H}'(\tilde{z})^T \tilde{M} \tilde{H}'(\tilde{z}).$$

Assume that $(\tilde{H}, \varphi)$ satisfies (6.2.11) and (6.2.12). Then (6.2.21) and (6.2.23) follow
by differentiation of (6.2.11) and (6.2.12), respectively, using $\frac{\partial^2}{\partial \tilde{z}_i \partial \tilde{z}_j} \varphi = \frac{\partial^2}{\partial \tilde{z}_j \partial \tilde{z}_i} \varphi$ and
$\frac{\partial^2}{\partial \tilde{z}_i \partial \tilde{z}_j} \varphi = \frac{\partial^2}{\partial \tilde{z}_j \partial \tilde{z}_i} \varphi$. Since $\mathcal{H}(\tilde{z})$ is independent of $\hat{z}$ we obtain by application of the
differential operator $\omega(\hat{z}) \cdot \nabla_{\tilde{z}}$ to (6.2.12) and differentiation of (6.2.11) with respect
to $\tilde{z}$ that

$$\mathcal{H}(\tilde{z}) \left(\omega(\hat{z}) \cdot \nabla_{\tilde{z}} \right) \hat{f}(0,\hat{z},\tilde{z}) = \left(\omega(\hat{z}) \cdot \nabla_{\tilde{z}} \right) \mathcal{H}(\tilde{z}) \hat{f}(0,\hat{z},\tilde{z})$$

$$= -\left(\omega(\hat{z}) \cdot \nabla_{\tilde{z}} \right) \nabla_{\tilde{z}} \varphi(\hat{z},\tilde{z}) = -\nabla_{\tilde{z}} (\omega(\hat{z}) \cdot \nabla_{\tilde{z}} \varphi(\hat{z},\tilde{z}))$$

$$= \nabla_{\tilde{z}} \left(\omega(\hat{z}) \cdot \hat{M} \hat{f}(0,\hat{z},\tilde{z}) \right),$$

which proves (6.2.22).

Conversely, assume that f and $\tilde{H}$ satisfy (6.2.21) – (6.2.23). Since by assumption the section $\tilde{D}_0(f, \hat{z}_0)$ is simply connected, the equation (6.2.23) implies that there exists a function $\Phi \in C^2(\tilde{D}_0(f, \hat{z}_0), \mathbb{R})$ with

$$- \mathcal{H}(\tilde{z})\hat{f}(0, \hat{z}_0, \tilde{z}) = \nabla_{\tilde{z}}\Phi(\tilde{z}) \tag{6.2.25}$$

for $\tilde{z} \in \tilde{D}_0(f, \hat{z}_0)$. Similarly, since $\hat{D}_0(f, \tilde{z})$ is Γ–star-shaped and therefore simply connected, equation (6.2.21) implies that for every $\tilde{z}$ there exists a function $\Psi_{\tilde{z}} \in C^2(\hat{D}_0(f, \tilde{z}), \mathbb{R})$ with

$$- \hat{M}\hat{f}(0, \hat{z}, \tilde{z}) = \nabla_{\hat{z}}\Psi_{\tilde{z}}(\hat{z}) . \tag{6.2.26}$$

By adding a suitable function only depending on $\tilde{z}$ we can normalize $\Psi_{\tilde{z}}$ such that $\Psi_{\tilde{z}}(\hat{z}_0) = 0$. Now define $\varphi : D_0(f) \to \mathbb{R}$ by

$$\varphi(\hat{z}, \tilde{z}) = \Psi_{\tilde{z}}(\hat{z}) + \Phi(\tilde{z}) .$$

Then

$$\varphi(\hat{z}_0, \tilde{z}) = \Phi(\tilde{z}) , \tag{6.2.27}$$

and (6.2.25) thus implies

$$- \mathcal{H}(\tilde{z})\hat{f}(0, \hat{z}_0, \tilde{z}) = \nabla_{\tilde{z}}\varphi(\hat{z}_0, \tilde{z}) , \tag{6.2.28}$$

whereas (6.2.26) implies

$$- \hat{M}\hat{f}(0, \hat{z}, \tilde{z}) = \nabla_{\hat{z}}\varphi(\hat{z}, \tilde{z}) \tag{6.2.29}$$

for all $(\hat{z}, \tilde{z}) \in D_0(f)$. This equation shows that φ satisfies (6.2.11).

For $\hat{z} \in \hat{D}_0(f, \tilde{z})$ let $\gamma : [0, s] \to \hat{D}_0(f, \tilde{z})$ be a path connecting $\hat{z}_0 = \gamma(0)$ with $\hat{z} = \gamma(s)$. Such a path exists, since we can choose for γ the integral curve of ω connecting $\hat{z}_0$ with $\hat{z}$. By smoothing this integral curve we can construct the path so that it is twice continuously differentiable. From (6.2.27) and (6.2.29) we then obtain

$$\varphi(\hat{z}, \tilde{z}) = \varphi(\gamma(s), \tilde{z})$$

$$= \int_0^s \frac{d}{d\sigma}\varphi(\gamma(\sigma), \tilde{z})d\sigma + \varphi(\gamma(0), \tilde{z})$$

$$= \int_0^s \gamma'(\sigma) \cdot \nabla_{\hat{z}}\varphi(\gamma(\sigma), \tilde{z})d\sigma + \varphi(\hat{z}_0, \tilde{z}) \tag{6.2.30}$$

$$= - \int_0^s \gamma'(\sigma) \cdot \hat{M}\hat{f}(0, \gamma(\sigma), \tilde{z})d\sigma + \Phi(\tilde{z}) .$$

Since by assumption $z \mapsto \hat{f}(0, z) \in C^2(D_0(f), \mathbb{R}^d)$ and since $\Phi \in C^2(\tilde{D}_0(f, \hat{z}_0), \mathbb{R})$, we conclude from this equation by the usual method that $\varphi \in C^2(D_0(f), \mathbb{R})$, varying the path of integration γ suitably.

Now choose γ in (6.2.30) to be the integral curve of the vector field ω. This integral curve satisfies

$$\gamma'(\sigma) = \omega(\gamma(\sigma)) .$$

We insert this equation into (6.2.30), differentiate the resulting equation with respect to $\tilde{z}$, and use (6.2.22), (6.2.27) and (6.2.28) to obtain

$$\nabla_{\tilde{z}}\varphi(\hat{z},\tilde{z}) = -\int_0^s \nabla_{\tilde{z}}\big(\omega(\gamma(\sigma))\cdot\hat{M}\,\tilde{f}(0,\gamma(\sigma),\tilde{z})\big)d\sigma + \nabla_{\tilde{z}}\varphi(\hat{z}_0,\tilde{z})$$

$$= -\int_0^s \mathcal{H}(\tilde{z})\,(\omega(\gamma(\sigma))\cdot\nabla_{\tilde{z}})\tilde{f}(0,\gamma(\sigma),\tilde{z})d\sigma - \mathcal{H}(\tilde{z})\tilde{f}(0,\hat{z}_0,\tilde{z})$$

$$= -\int_0^s (\omega(\gamma(\sigma))\cdot\nabla_{\tilde{z}})\mathcal{H}(\tilde{z})\tilde{f}(0,\gamma(\sigma),\tilde{z})d\sigma - \mathcal{H}(\tilde{z})\tilde{f}(0,\hat{z}_0,\tilde{z})$$

$$= -\int_0^s \gamma'(\sigma)\cdot\nabla\big(\mathcal{H}(\tilde{z})\tilde{f}(0,\gamma(\sigma),\tilde{z})\big)d\sigma - \mathcal{H}(\tilde{z})\tilde{f}(0,\hat{z}_0,\tilde{z})$$

$$= -\int_0^s \frac{d}{d\sigma}\big(\mathcal{H}(\tilde{z})\tilde{f}(0,\gamma(\sigma),\tilde{z})\big)d\sigma - \mathcal{H}(\tilde{z})\tilde{f}(0,\gamma(0),\tilde{z})$$

$$= -\mathcal{H}(\tilde{z})\tilde{f}(0,\gamma(s),\tilde{z}) = -\mathcal{H}(\tilde{z})\tilde{f}(0,\hat{z},\tilde{z}).$$

This shows that (6.2.12) is fulfilled. The proof of (I) is complete.

(II) Let M and $\tilde{H} : \mathbb{R}^{N-d} \to \mathbb{R}^{N-d}$ satisfy (iii) - (vi). The mapping obtained by transformation of $\tilde{H}$ to the original coordinates is given by $\tilde{J}^{-1} \circ \tilde{H} \circ \tilde{J} : \ker(B) \to \ker(B) \subseteq \mathbb{R}^N$. We denote this mapping again by $\tilde{H}$ and define the transformation field H by

$$H(z) = Pz + \tilde{H}(Qz). \tag{6.2.31}$$

From (v), (vi) and Theorem 6.1.2 it follows that the pair obtained by transformation of (f,B) with this transformation field is of pre-monotone type. Moreover, in the proof of Theorem 6.1.2 we showed that for constitutive equations of the form (6.1.2) -(6.1.4) with $\overline{L}$, $\hat{L}$, M satisfying (v), and for transformation fields of the form (6.2.31) the condition (iv) of Theorem 5.3.1 holds. Using these two results, we infer from Theorem 5.4.1 that the pair obtained by transformation of (f,B) with the vector field H defined in (6.2.31) is of gradient type and consequently belongs to the subclass $\mathcal{G}$ of $\mathcal{M}^*$ if and only if H satisfies the partial differential equation (6.2.20). Part I of this proof shows that this H satisfies (6.2.20) if and only if (iii) and (iv) are fulfilled. Together this means that the vector field H from (6.2.31) transforms (f,B) to the class $\mathcal{G}$, hence $(f,B) \in \mathcal{T}\mathcal{G}$.

(III) It suffices to show that if $(f,B) \in \mathcal{T}\mathcal{G}$, then M and $\tilde{H}$ exist satisfying conditions (iii) - (vi), since the converse was just proved. Since $\mathcal{T}\mathcal{G} \subseteq \mathcal{T}\mathcal{M}^*$, it follows that if (f,B) is transformed to gradient type by the transformation field H, then H also transforms (f,B) to pre-monotone type. From Corollary 6.1.4 we thus conclude that M exists satisfying conditions (v) and (vi), and that H must be of the form (6.1.47) for a bijective linear mapping A and for all $z \in D_0(f)+R(\hat{L}^{-1}\overline{L})$. From Lemma 5.2.2 we also infer that the vector field $G(z) = AH(z)$, which on $D_0(f) + R(\hat{L}^{-1}\overline{L})$ is of the form

$$G(z) = AH(z) = Pz + \tilde{H}(Qz),$$

transforms (f,B) to gradient type. From Theorem 5.4.1 it thus follows that G must satisfy the partial differential equation (6.2.20), and part I of this proof thus implies that M and $\tilde{H}$ fulfill conditions (iii) and (iv). The proof of the theorem is complete.

Remark. Equation (6.2.21) is a compatibility condition for $\hat{f}$. Additional compatibility conditions for $\hat{f}$ and $\tilde{f}$ are implicitely contained in (6.2.22) and (6.2.23), since, firstly, unknown in (6.2.22), (6.2.23) are the $N - d$ component functions $\tilde{H}_1, \ldots, \tilde{H}_{N-d}$ of $\tilde{H}$, whereas (6.2.22) consists of $N - d$ equations and (6.2.23) of $(N - d)(N - d - 1)/2$ equations. If $N - d > 1$, the number of equations is therefore larger than the number of unknowns. Secondly, the unknown vector field $\tilde{H}$ only depends on $\tilde{z}$, while, as in the system (5.1.11), the coefficient functions $\hat{f}$ and $\tilde{f}$ in the system of partial differential equations (6.2.22) depend on the additional parameter $\hat{z}$, and the system must be satisfied for all values $(\hat{z}, \tilde{z}) \in D_0(f)$. This can only be the case if $\hat{f}$ and $\tilde{f}$ satisfy suitable compatibility conditions.

The case $N - d = 1$, where the system (6.2.23) is not present, is of practical importance, since $N - d$ is normally equal to the number of hardening parameters, and many simpler constitutive models contain just one hardening parameter. In this case (6.2.22) reduces to an ordinary differential equation. The model of Bodner and Partom, which will be discussed in Chapter 8, belongs to this class of models.

If $N - d > 1$, it is sometimes possible to choose $\hat{z}_0$ such that

$$\tilde{f}(0, \hat{z}_0, \tilde{z}) = 0,$$

for all $\tilde{z} \in \tilde{D}_0(f, \hat{z}_0)$. Also in this case (6.2.23) is automatically satisfied. The rate independent constitutive relations discussed in Chapter 7 and in particular the constitutive relation discussed in Example 2 of that chapter are of this form.

An **example** of the application of Theorem 6.2.4 is given in Chapter 8. However, because the investigation in Chapter 7 of rate independent constitutive equations runs in parallel to the investigation of rate dependent problems in this section and is somewhat simpler, the two examples presented in Sections 7.3 and 7.4 are also very helpful for understanding Theorem 6.2.4. Therefore the reader is advised to study those examples first, though Theorem 6.2.4 is not directly applied in these examples.

Chapter 7

Transformation of Rate Independent Constitutive Equations

In this chapter we study the transformation of rate independent constitutive equations. The general theory is developed in Sections 7.1 and 7.2. This theory runs in parallel to the theory for rate dependent constitutive equations developed in Chapters 5 and 6, but since rate independent constitutive equations are strongly structured, as is shown by Lemma 2.1.2, the considerations are slightly simpler. An important difference to the theory for rate dependent constitutive equations is that rate independent equations of monotone type are necessarily of monotone-gradient type, and transformation to monotone type therefore amounts to transformation to gradient type. As is shown in Theorem 7.2.1, the transformation field must consequently satisfy a system of partial differential equations analogous to the system of partial differential equations for the rate dependent case stated in Theorem 5.4.1. It will be seen that in a certain sense, the theory amounts to the transformation of the yield surface in the rate independent case. The application of the theory is demonstrated in Sections 7.3 and 7.4 by two examples.

7.1 Transformation of Constitutive Equations Containing Set-Valued Operators

According to Lemma 2.1.2, the constitutive equation in the system

$$\rho u_{tt} = \operatorname{div}_x T \tag{7.1.1}$$

$$T = D\big(\varepsilon(\nabla_x u) - Bz\big) \tag{7.1.2}$$

$$z_t \in f\big(\varepsilon(\nabla_x u), z\big) \tag{7.1.3}$$

is rate independent if the set-valued function

$$f : \mathcal{S}^3 \times \mathbb{R}^N \to \mathcal{P}(\mathbb{R}^N)$$

satisfies

$$f(\varepsilon, z) = \bigcup_{\lambda \geq 0} \lambda f(\varepsilon, z)$$

for all $z \in \mathbb{R}^N$, which means that $f(\varepsilon, z)$ is a conic subset of $\mathbb{R}^N$ for all $z \in \mathbb{R}^N$. In the investigations of this chapter the function f is of the form

$$f(\varepsilon, z) = g_K(\overline{L}\varepsilon - \hat{L}z). \tag{7.1.4}$$

Here $\overline{L} : \mathcal{S}^3 \to \mathbb{R}^N$ is a linear mapping, $\hat{L} : \mathbb{R}^N \to \mathbb{R}^N$ is an invertible linear mapping and $K \neq \emptyset$ is a closed convex subset of $\mathbb{R}^N$. The function $g_K : \mathbb{R}^N \to \mathcal{P}(\mathbb{R}^N)$ is required to satisfy

$$g_K(z) = \bigcup_{\lambda \geq 0} \lambda g_K(z) \tag{7.1.5}$$

for all $z \in \mathbb{R}^N$ and

$$g_K(z) \begin{cases} = \{0\}, & z \in \overset{\circ}{K} \\ \neq \{0\}, \neq \emptyset, & z \in \partial K \\ = \emptyset, & z \in \mathbb{R}^N \backslash K. \end{cases} \tag{7.1.6}$$

Because of (7.1.5), for functions f of this form the constitutive equations are rate independent. The boundary ∂K of K and the mappings $\overline{L}, \hat{L}$ are chosen such that

$$\{(\varepsilon, z) \in \mathcal{S}^3 \times \mathbb{R}^N \mid \overline{L}\varepsilon - \hat{L}z \in \partial K\}$$

is a hypersurface in $\mathcal{S}^3 \times \mathbb{R}^N$, and therefore so is the image of this surface under the mapping $(\varepsilon, z) \mapsto (T, z) = \big(D(\varepsilon - Bz), z\big)$. This image is called the yield surface. If (T, z) belongs to the yield surface, then (7.1.6) implies that $g_K(\overline{L}\varepsilon - \hat{L}z)$ differs from $\{0\}$ and from $\emptyset$, and consequently the constitutive equations (7.1.3), (7.1.4) allow z_t to be different from zero. This means that the material displays plastic flow for (T, z) on the yield surface.

The relations (7.1.5) and (7.1.6) are satisfied by the subdifferential $\partial \chi_K$ of the characteristic function $\chi_K : \mathbb{R}^N \to [0, \infty]$ of K, which is defined by

$$\chi_K(z) = \begin{cases} 0, & z \in K \\ \infty, & z \in \mathbb{R}^N \backslash K. \end{cases}$$

Since we assumed that K is a convex set, it follows that the characteristic function is convex, and that the subdifferential $\partial \chi_K : \mathbb{R}^N \to \mathcal{P}(\mathbb{R}^N)$ is a maximal monotone operator, cf. [23, p.25]. Moreover, it is easy to show that a function $g_K : \mathbb{R}^N \to \mathcal{P}(\mathbb{R}^N)$ which satisfies (7.1.5) and (7.1.6), is monotone if and only if

$$g_K(z) \subseteq \partial \chi_K(z)$$

holds for all $z \in \mathbb{R}^N$. Consequently, g_K is maximal monotone if and only if $g_K = \partial \chi_K$. If ∂K is smooth in a neighborhood of $z \in \partial K$, then

$$\partial \chi_K(z) = \{\lambda n(z) \mid \lambda \geq 0\},$$

where $n(z)$ is the exterior unit normal vector to ∂K at z. However, in engineering models generically the ray $g_K(z)$ does not have the direction of $n(z)$, hence $g_K(z) \not\subseteq \partial \chi_K(z)$, and g_K is not monotone. Transformation to monotone type therefore essentially means transforming the yield surface ∂K to a new surface ∂K_H and transforming the set-valued function g_K to the subdifferential of the characteristic function of the set with boundary ∂K_H.

We first examine the transformation of interior variables for the system (7.1.1) – (7.1.3) with a general set-valued function $f : S^3 \times \mathbb{R}^N \to \mathcal{P}(\mathbb{R}^N)$. For a transformation field $H : \mathbb{R}^N \to \mathbb{R}^N$ and for a set $X \subseteq \mathbb{R}^N$ we denote by

$$H'(z)X = \{[H'(z)]w \mid w \in X\} \subseteq \mathbb{R}^N$$

the image of X under the linear mapping $H'(z) : \mathbb{R}^N \to \mathbb{R}^N$.

Lemma 7.1.1 *If (u, z) is a solution of (7.1.1)-(7.1.3), then the function (u, h) with the transformed interior variables $h(x, t) = H(z(x, t))$ satisfies*

$$\rho u_{tt} = \mathrm{div}_x T$$
$$T = D(\varepsilon(\nabla_x u) - BH^{-1}(h))$$
$$h_t \in f_H(\varepsilon(\nabla_x u), h),$$

where

$$f_H(\varepsilon, h) = H'(H^{-1}(h))f(\varepsilon, H^{-1}(h))$$
$$= \{H'(H^{-1}(h))w \mid w \in f(\varepsilon, H^{-1}(h))\}.$$

Proof. We have

$$\frac{\partial}{\partial t}h(x, t) = \frac{\partial}{\partial t}H(z(x, t))$$
$$= H'(z(x, t))\frac{\partial}{\partial t}z(x, t) \in H'(z(x, t))f(\varepsilon(x, t), z(x, t))$$
$$= H'\left(H^{-1}(h(x, t))\right)f\left(\varepsilon(x, t), H^{-1}(h(x, t))\right)$$
$$= f_H(\varepsilon(x, t), h(x, t)).$$

Now we need a technical lemma:

Lemma 7.1.2 *Assume that $K \neq \emptyset$ is a closed, convex subset of $\mathbb{R}^N$ and that the image $H(K)$ of K under the transformation field $H : \mathbb{R}^N \to \mathbb{R}^N$ is convex. Let $\chi_{H(K)}$ be the characteristic function of $H(K)$. Then $\chi_K = \chi_{H(K)} \circ H$ is the characteristic function of K, and the subdifferentials satisfy*

$$\partial\chi_K(z) = H'(z)^T \partial\chi_{H(K)}(H(z))$$

for all $z \in \mathbb{R}^N$.

Proof. It is obvious that $\chi_K = \chi_{H(K)} \circ H$ is the characteristic function of K. To prove the statement about the subdifferentials it suffices to show that

$$H'(z)^T \partial\chi_{H(K)}(H(z)) \subseteq \partial\chi_K(z), \tag{7.1.7}$$

since by the symmetry of our assumptions we then obtain from $\chi_{H(K)} = \chi_K \circ H^{-1}$ that also

$$(H^{-1})'(\bar{z})^T \partial\chi_K(H^{-1}(\bar{z})) \subseteq \partial\chi_{H(K)}(\bar{z})$$

holds. If we choose $\overline{z} = H(z)$ in this relation, we obtain

$$\partial\chi_K(z) \subseteq \left[(H^{-1})'(H(z))\right]^{-T} \partial\chi_{H(K)}(H(z)) = H'(z)^T \partial\chi_{H(K)}(H(z)),$$

which together with (7.1.7) proves the statement of the lemma. To prove (7.1.7), note that if $z \notin K$ then $H(z) \notin H(K)$, hence $\partial\chi_K(z) = \partial\chi_{H(K)}(H(z)) = \emptyset$, which shows that in this case (7.1.7) is satisfied. Now let $z \in K$. Then $w \in \mathbb{R}^N$ belongs to $\partial\chi_{H(K)}(H(z))$ if and only if

$$\chi_{H(K)}(\overline{z}) \geq \chi_{H(K)}(H(z)) + w \cdot (\overline{z} - H(z)) \tag{7.1.8}$$

for all $\overline{z} \in \mathbb{R}^N$, and $[H'(z)^T]w$ belongs to $\partial\chi_K(z)$ if and only if

$$\chi_K(\overline{z}) \geq \chi_K(z) + (H'(z)^T w) \cdot (\overline{z} - z) \tag{7.1.9}$$

for all $\overline{z} \in \mathbb{R}^N$. Therefore (7.1.7) follows if we show that (7.1.8) implies (7.1.9). Since $\chi_K(z) = 0$ and $\chi_K(\overline{z}) = \infty$ for $\overline{z} \notin K$, (7.1.9) certainly holds for $\overline{z} \notin K$. Therefore assume that $\overline{z} \in K$. The convexity of K then yields $(t\overline{z} + (1 - t)z) \in K$ for all $0 \leq t \leq 1$. If we replace $\overline{z}$ in (7.1.8) by $H(t\overline{z} + (1 - t)z)$ and note that $\chi_{H(K)}(H(z)) = \chi_{H(K)}\left(H(t\overline{z} + (1 - t)z)\right) = 0$, we obtain from this relation

$$0 \geq w \cdot \left[H(t\overline{z} + (1 - t)z) - H(z)\right] = w \cdot H'(z)(\overline{z} - z)t + o(t), \quad t \searrow 0,$$

hence

$$w \cdot H'(z)(\overline{z} - z) = (H'(z)^T w) \cdot (\overline{z} - z) \leq 0,$$

from which we deduce (7.1.9) using $\chi_K(\overline{z}) = \chi_K(z) = 0$. This completes the proof of the lemma.

7.2   Transformation to Monotone-Gradient Type

Now we can study the transformation of the constitutive equations (7.1.2), (7.1.3) to constitutive equations of monotone-gradient type. We assume that the set valued function f is of the form (7.1.4) with g_K satisfying (7.1.5), (7.1.6). If the transformed constitutive equations are of the form

$$T = D(\varepsilon - BH^{-1}(h))$$

$$h_t \in f_H(\varepsilon, h) = \partial\chi_{K^*}(-\rho\nabla_h\psi(\varepsilon, h))$$

for a suitable free energy ψ, for which the symmetric linear mapping $M : \mathbb{R}^N \to \mathbb{R}^N$ defined by

$$Mh = \rho\nabla_h\psi(0, h)$$

is positive definite, and for the characteristic function χ_{K^*} of a closed convex subset K^* of $\mathbb{R}^N$, then according to Lemma 7.1.1 the transformation field H must satisfy the equation

$$\partial\chi_{K^*}(-\rho\nabla_h\psi(\varepsilon, h)) = H'(H^{-1}(h))f(\varepsilon, H^{-1}(h)) \tag{7.2.1}$$

for all $(\varepsilon, h) \in \mathcal{S}^3 \times \mathbb{R}^N$. For $\varepsilon = 0$ we obtain from (7.2.1) the equation

$$\partial \chi_{K^*} \big(- MH(z) \big) = H'(z) f(0, z) = H'(z) g_K(-\hat{L}z), \qquad (7.2.2)$$

which must be satisfied for all $z \in \mathbb{R}^N$. Since, as always, we assume that $H'(z)$ is an invertible linear mapping, we deduce from this equation and from (7.1.6) that $-MH(z) \in \partial K^*$ if and only if $-\hat{L}z \in \partial K$ and $-MH(z) \in \overset{\circ}{K}{}^*$ if and only if $-\hat{L}z \in \overset{\circ}{K}$, hence

$$K^* = -MH(-\hat{L}^{-1}K).$$

By our assumptions, K and K^* are convex, and since $-\hat{L}^{-1}$ is linear, we infer from this equation that the set

$$(-MH)^{-1} K^* = -\hat{L}^{-1} K$$

is also convex. Consequently Lemma 7.1.2 implies

$$(-MH)'(z)^T \partial \chi_{K^*}(-MH(z)) = \partial \chi_{(-MH)^{-1}K^*}(z) = \partial \chi_{-\hat{L}^{-1}K}(z).$$

With this relation we deduce from (7.2.2) the equivalent equation

$$\partial \chi_{-\hat{L}^{-1}K}(z) = -H'(z)^T MH'(z) g_K(-\hat{L}z), \quad z \in \mathbb{R}^N. \qquad (7.2.3)$$

This is a partial differential equation for H, since the only unknowns are H and M. As in Theorem 6.1.2, we can show that if H is of the special form (6.1.14), then the equations (7.2.1) and (7.2.2) are also equivalent. This means that such an H transforms the given constitutive equations to monotone-gradient type if and only if they are solutions of the partial differential equation (7.2.3). The precise result is:

Theorem 7.2.1 *Assume that a symmetric, positive definite $N \times N$-matrix M exists with*

(i) $\qquad M^{-1} B^T D = \hat{L}^{-1} \overline{L}$

(ii) $\qquad L = M - B^T DB$ *is positive semi-definite.*

Moreover, assume that the transformation field H is of the form

$$H(z) = Pz + \tilde{H}(Qz), \qquad (7.2.4)$$

where $P : \mathbb{R}^N \to \mathbb{R}^N$ is the projection along $\ker(B)$ onto the range $R(\hat{L}^{-1}\overline{L})$, where $Q = I - P$ is the projection along $R(\hat{L}^{-1}\overline{L})$ onto $\ker(B)$, and where $\tilde{H} : \ker(B) \to \ker(B) \subseteq \mathbb{R}^N$. Assume also that the sets K from g_K and

$$K^* = -MH(-\hat{L}^{-1}K)$$

are convex. Then H transforms the constitutive equations (7.1.2) – (7.1.4) to constitutive equations of gradient type with free energy ψ satisfying $Mh = \rho \nabla_h \psi(0, h)$, if and only if H solves the system of partial differential equations

$$- H'(z)^T MH'(z) g_K(-\hat{L}z) = \partial \chi_{-\hat{L}^{-1}K}(z) \qquad (7.2.5)$$

for all $z \in \mathbb{R}^N$. The transformed equations are of the form

$$T = D(\varepsilon - Bh) \tag{7.2.6}$$
$$h_t \in \partial\chi_{K^*}(-\rho\nabla_h\psi(\varepsilon, h)) \tag{7.2.7}$$

with

$$\rho\psi(\varepsilon, h) = \frac{1}{2}[D(\varepsilon - Bh)] \cdot (\varepsilon - Bh) + \frac{1}{2}(Lh) \cdot h,$$

$$\partial\chi_{K^*}(h) = H'(H^{-1}(-M^{-1}h))g_K(-\hat{L}H^{-1}(-M^{-1}h)).$$

If $0 \in K^*$, then the equations (7.2.6), (7.2.7) are of monotone-gradient type.

Proof. Exactly as in the proof of Theorem 6.1.2 it is shown that if H satisfies (7.2.4) then the transformed constitutive equations are of the form

$$T = D(\varepsilon - Bh)$$
$$h_t = g_H(-\rho\nabla_h\psi(\varepsilon, h))$$

with g_H satisfying

$$g_H(-\rho\nabla_h\psi(\varepsilon, h)) = H'(H^{-1}(h))f(\varepsilon, H^{-1}(h))$$

and therefore

$$g_H(-MH(z)) = H'(z)f(0, z) = H'(z)g_K(-\hat{L}z),$$

for all $z \in \mathbb{R}^N$. This equation and (7.2.2) both hold at the same time if and only if $g_H = \partial\chi_{K^*}$. We remarked above that the equations (7.2.2) and (7.2.3) are equivalent. Whence the equations (7.2.2) and (7.2.5) are equivalent, which shows that $g_H = \partial\chi_{K^*}$ if and only if (7.2.5) holds.

If $0 \in K^*$, then $0 \in \partial\chi_{K^*}(0)$ and χ_{K^*} is a proper convex function. Consequently, the subdifferential $\partial\chi_{K^*}$ is monotone. Using the remark after Definition 3.1.1 we thus see by comparision with this definition that the equations (7.2.6), (7.2.7) are of monotone-gradient type. This proves the theorem.

As with the system (6.2.20) for the rate dependent case, the system of partial differential equations (7.2.5) has solutions of the form (7.2.4) only if the coefficient function g_K satisfies compatibility conditions. To derive these compatibility conditions and to derive partial differential equations for $\tilde{H}$, note that for $z \in \mathbb{R}^N \backslash \partial(-\hat{L}^{-1}K)$ equation (7.2.5) is satisfied for every vector field H, since by definition of the subdifferential and by (7.1.6), for $z \in (-\hat{L}^{-1}K)^0$ the equality $g_K(-\hat{L}z) = \{0\} = \partial\chi_{-\hat{L}^{-1}K}(z)$ holds and for $z \in \mathbb{R}^N \backslash (-\hat{L}^{-1}K)$ the equation $g_K(-\hat{L}z) = \emptyset = \partial\chi_{-\hat{L}^{-1}K}(z)$ is fulfilled.

Therefore it suffies to determine H such that (7.2.5) holds on $\partial(-\hat{L}^{-1}K)$. As always, for the transformation field H we require that the Jacobi matrix $H'(z)$ is invertible for every z. For such a vector field, equation (7.2.5) is satisfied if and only if for every $z \in \partial(-\hat{L}^{-1}K)$ the conic subset $g_K(-\hat{L}z)$ is mapped in a one-to-one way onto the conic subset $\partial\chi_{-\hat{L}^{-1}K}(z)$. This means in particular that for every element $\gamma(z) \in g_K(-\hat{L}z)$ with $\gamma(z) \neq 0$ there exists a vector $n(z) \in \partial\chi_{-\hat{L}^{-1}K}(z)$ with $|n(z)| = 1$ and a $\lambda(z) > 0$ with

$$- H'(z)^T M H'(z)\gamma(z) = \lambda(z)n(z). \tag{7.2.8}$$

As in Lemma 6.2.1 we regard $\tilde{H}$ as a map $\tilde{H} : \ker(B) \to \ker(B)$. With the orthogonal projections $P^\perp : \mathbb{R}^N \to \mathbb{R}^N$ onto the subspace $R(\hat{L}^{-1}\overline{L})$ and $Q^\perp : \mathbb{R}^N \to \mathbb{R}^N$ onto the subspace $\ker(B)$, and with the mappings $\hat{M} = P^\perp M\big|_{R(\hat{L}^{-1}\overline{L})}$, $\tilde{M} = Q^\perp M\big|_{\ker(B)}$ introduced before Lemma 6.2.1, we then obtain

Lemma 7.2.2 *If H is of the form*

$$H(z) = Pz + \tilde{H}(Qz),$$

then (7.2.8) is satisfied if and only if

$$-\hat{M}P\gamma(z) = \lambda(z)P^\perp n(z) \tag{7.2.9}$$

$$-Q^\perp \tilde{H}'(Qz)^T \tilde{M} \tilde{H}'(Qz)Q\gamma(z) = \lambda(z)Q^\perp n(z). \tag{7.2.10}$$

The **proof** is exactly the same as the proof of Lemma 6.2.1.

We now use the notation

$$\tilde{z} = Qz, \quad \hat{\gamma}(z) = P\gamma(z), \quad \tilde{\gamma}(z) = Q\gamma(z).$$

Since $\tilde{H}'(\tilde{z}) : \ker(B) \to \ker(B)$, the operator $Q^\perp$ on the left hand side of (7.2.10) is the injection of $\ker(B)$ into $\mathbb{R}^N$. In the following we identify elements of $\ker(B)$ with elements of $\mathbb{R}^N$ and thus drop $Q^\perp$ on the left hand side of this equation.

Corollary 7.2.3 *Assume that the transformation field is of the form*

$$H(z) = Pz + \tilde{H}(Qz).$$

Then (7.2.5) is satisfield if and only if the following two conditions are satisfied:

(i) *For every $z \in \partial(-\hat{L}^{-1}K)$ and every $\gamma(z) \in g_K(-\hat{L}z)$ with $\gamma(z) \neq 0$ there exist a unit vector $n(z) \in \partial\chi_{-\hat{L}^{-1}K}(z)$ and a positive number $\lambda(z)$ with*

$$-\hat{M}\hat{\gamma}(z) = \lambda(z) \cdot P^\perp n(z) \tag{7.2.11}$$

$$-\tilde{H}'(\tilde{z})^T \tilde{M} \tilde{H}'(\tilde{z})\tilde{\gamma}(z) = \begin{cases} Q^\perp n(z)\dfrac{|\hat{M}\hat{\gamma}(z)|}{|P^\perp n(z)|}, & \text{if } P^\perp n(z) \neq 0 \\ Q^\perp n(z)\lambda(z), & \text{if } P^\perp n(z) = 0. \end{cases} \tag{7.2.12}$$

(ii) *The mapping $\gamma(z) \to n(z)$ thus defined is onto the set*

$$\{n(z) \in \partial\chi_{-\hat{L}^{-1}K}(z) \mid n(z) = 1\}.$$

Proof. For $P^{\perp}n(z) \neq 0$ the equation (7.2.11) implies $\lambda(z) = |\hat{M}\hat{\gamma}(z)|/|P^{\perp}n(z)|$, which shows that the equations (7.2.11), (7.2.12) coincide with (7.2.9), (7.2.10). Whence they are equivalent to (7.2.8). Consequently, statement (i) implies that

$$-H'(z)^T M H'(z) g_K(-\hat{L}z) \subseteq \partial \chi_{-\hat{L}^{-1}K}(z)$$

for every $z \in \partial(-\hat{L}^{-1}K)$, and statement (ii) implies that in this relation equality holds. Therefore (7.2.5) holds on $\partial(-\hat{L}^{-1}K)$ and thus, as remarked above, on all of $\mathbb{R}^N$. This proves the corollary.

If in a neighborhood U of $z_0 \in \partial(-\hat{L}^{-1}K)$ the boundary $\partial(-\hat{L}^{-1}K)$ is a smooth $(N-1)$–dimensional surface, then for every $z \in U \cap \partial(-\hat{L}^{-1}K)$ the set $\partial \chi_{-\hat{L}^{-1}K}(z)$ is a one-dimensional ray

$$\partial \chi_{-\hat{L}^{-1}K}(z) = \{\lambda n(z) \mid \lambda \geq 0\},$$

where $n(z)$ is the unit normal vector to $\partial(-\hat{L}^{-1}K)$ pointing into the exterior of $-\hat{L}^{-1}K$. Consequently, if (7.2.5) is satisfied,

$$g_K(-\hat{L}z) = \{\lambda\gamma(z) \mid \lambda \geq 0\}$$

is also a one-dimensional ray, where $\gamma : U \cap \partial(-\hat{L}^{-1}K) \to g_K(-\hat{L}z) \subseteq \mathbb{R}^N$ with $\gamma(z) \neq 0$ is an arbitrarily chosen function. If $P^{\perp}n(z) \neq 0$ for all $z \in U \cap \partial(-\hat{L}^{-1}K)$, that is, if the subspace $R(\hat{L}^{-1}\overline{L})$ is nowhere tangential to $U \cap \partial(-\hat{L}^{-1}K)$, then with this function γ inserted, equation (7.2.12) becomes a system of $N - d$ first order partial differential equations (one ordinary differential equation if $N - d = 1$) for the $N - d$ components of $\tilde{H}$ on $U \cap \partial(-\hat{L}^{-1}K)$. This system can be used to determine $\tilde{H}(\tilde{z})$ for $\tilde{z}$ in this neighborhood of $\tilde{z}_0$. We illustrate this in the following two examples, where we derive the ordinary or partial differential equations for $\tilde{H}$. However, the derivation given below is directly based on the partial differential equation (7.2.5) from Theorem 7.2.1, not on equation (7.2.12), since for our examples this method of derivation is simpler.

7.3   Example 1: One Variable of Isotropic Hardening

This example is obtained from the Prandtl-Reuss model (Example 1 in Section 2.2) by adding a parameter of isotropic hardening. For this model the equations (7.1.1) – (7.1.3) have the form

$$\rho u_{tt} = \mathrm{div}_x T \tag{7.3.1}$$

$$T = D\big(\varepsilon(\nabla_x u) - \varepsilon_p\big) \tag{7.3.2}$$

$$\frac{\partial}{\partial t}\varepsilon_p \in \partial\chi\left(\frac{|P_0 T|}{\zeta}\right)\frac{P_0 T}{|P_0 T|} \tag{7.3.3}$$

$$\frac{\partial}{\partial t}\zeta = \left|\frac{\partial}{\partial t}\varepsilon_p\right| k(|P_0 T|, \zeta), \tag{7.3.4}$$

where $P_0 T = T - \frac{1}{3} \text{trace}(T)I$ is the stress deviator, $\varepsilon_p(x, t) \in \mathcal{S}^3$ is the plastic strain tensor, $\zeta(x, t) > 0$ is the hardening parameter, and $\partial \chi$ is the subdifferential of the characteristic function $\chi : \mathbb{R} \to [0, \infty]$ of the interval $(-\infty, 1]$, which is defined by

$$\chi(r) = \begin{cases} 0 & , r \leq 1 \\ +\infty & , r > 1. \end{cases}$$

This subdifferential is given by

$$\partial \chi(r) = \begin{cases} \{0\} & , r < 1 \\ [0, \infty) & , r = 1 \\ \emptyset & , r > 1. \end{cases}$$

For $z \in \mathbb{R}^N$ and $\Lambda \subseteq \mathbb{R}$ we set

$$\Lambda z = \{\lambda z \mid \lambda \in \Lambda\}.$$

This notation is used in (7.3.3). Finally, $k = [0, \infty) \times (0, \infty) \to (0, \infty)$ is a given function determining the evolution of the hardening parameter and therefore the hardening behavior of the material.

The initial conditions for $(u(x, t), \varepsilon_p(x, t), \zeta(x, t))$ at $t = 0$ are chosen such that $\zeta(x, 0) \geq \zeta_0$ for all x and a positive constant ζ_0. Since the right hand side of (7.3.4) is non-negative, it follows that $\zeta(x, t) \geq \zeta_0$ for all (x, t). Therefore the form of the constitutive equations (7.3.3), (7.3.4) for $\zeta < \zeta_0$ is not important for the solution, and in the following investigations we can define the constitutive equations for $\zeta \leq 0$ in a way which reduces the technical difficulties of the investigations.

The constitutive equations (7.3.2) – (7.3.4) are of pre-monotone type. To show this, we use the isomorphism identifying the spaces $\mathcal{S}^3$ and $\mathbb{R}^6$ defined by (3.3.1). We set $N = 7$ and choose $z = (\varepsilon_p, \zeta) \in \mathcal{S}^3 \times \mathbb{R} \cong \mathbb{R}^N$ for the vector of internal variables. The map $B : \mathcal{S}^3 \times \mathbb{R} \to \mathcal{S}^3$ is defined by

$$Bz = B(\hat{z}, \eta) = \hat{z} \in \mathcal{S}^3. \tag{7.3.5}$$

The transpose $B^T : \mathcal{S}^3 \to \mathcal{S}^3 \times \mathbb{R}$ is given by $B^T \hat{z} = (\hat{z}, 0)$, hence $B^T DB z = B^T DB(\hat{z}, \eta) = (D\hat{z}, 0)$ and $B^T D\varepsilon = (D\varepsilon, 0)$. We define the linear symmetric mapping $\hat{L} : \mathcal{S}^3 \times \mathbb{R} \to \mathcal{S}^3 \times \mathbb{R}$ by

$$\hat{L}z = \hat{L}(\hat{z}, \eta) = (D\hat{z}, \eta). \tag{7.3.6}$$

This mapping is positive definite since $D : \mathcal{S}^3 \to \mathcal{S}^3$ is positive definite, and

$$L = \hat{L} - B^T DB,$$

which satisfies $Lz = L(\hat{z}, \eta) = (0, \eta)$, is positive semi-definite. If we now define $g : \mathcal{S}^3 \times \mathbb{R}^- \to \mathcal{P}(\mathcal{S}^3 \times \mathbb{R})$ by

$$g(z) = g(\hat{z}, \eta) = \partial \chi \left(\frac{|P_0 \hat{z}|}{-\eta} \right) \left(\hat{g}(\hat{z}), k(|P_0 \hat{z}|, -\eta) \right) \subseteq \mathcal{S}^3 \times \mathbb{R}, \tag{7.3.7}$$

where

$$\hat{g}(\hat{z}) = \frac{P_0\hat{z}}{|P_0\hat{z}|},$$

then the equations (7.3.1) – (7.3.4) can be written in the form

$$\rho u_{tt} = \text{div}_x T \qquad (7.3.8)$$

$$T = D(\varepsilon - Bz) \qquad (7.3.9)$$

$$z_t \in g(\overline{L}\varepsilon - \hat{L}z) = g(-\rho\nabla_z\psi(\varepsilon, z)) \qquad (7.3.10)$$

with $\overline{L} = B^T D$ and with

$$\rho\psi(\varepsilon, z) = \frac{1}{2}\Big[D(\varepsilon - Bz)\Big] \cdot (\varepsilon - Bz) + \frac{1}{2}(Lz) \cdot z \qquad (7.3.11)$$

$$= \frac{1}{2}\Big[D(\varepsilon - \hat{z})\Big] \cdot (\varepsilon - \hat{z}) + \frac{1}{2}\eta^2, \quad z = (\hat{z}, \eta).$$

Since (7.3.9) and (7.3.10) satisfy the conditions of Definition 3.1.1, the constitutive equations (7.3.2) – (7.3.4) are of pre-monotone type.

(7.3.7) implies that $g(z)$ differs from $\{0\}$ and from $\emptyset$ only if z belongs to the boundary of the convex set

$$K = \Big\{z = (\hat{z}, \eta) \,\Big|\, |P_0\hat{z}| \leq -\eta\Big\} \subseteq S^3 \times \mathbb{R}_0^-. \qquad (7.3.12)$$

Since $0 \in K$, it follows that the constitutive equations (7.3.2) – (7.3.4) are of monotone-gradient type if and only if we can extend g to all of $S^3 \times \mathbb{R}$ such that the extended function is equal to the subdifferential $\partial\chi_K$ of the characteristic function χ_K of K.

Lemma 7.3.1 *For the set K defined in (7.3.12) and for $z = (\hat{z}, \eta) \in S^3 \times \mathbb{R}$ with $\eta \neq 0$ we have*

$$\partial\chi_K(z) = \partial\chi\Big(\frac{|P_0\hat{z}|}{-\eta}\Big)\big(\hat{g}(\hat{z}), 1\big).$$

We leave the proof to the reader.

If we extend the function g to $z \in S^3 \times \mathbb{R}_0^+$ by $g(z) = \partial\chi_K(z)$, it follows from this lemma and from (7.3.7) immediately that $g = \partial\chi_K$ if and only if

$$k(\eta, \eta) = 1 \qquad (7.3.13)$$

for all $\eta > 0$, since $\partial\chi(\frac{|P_0\hat{z}|}{-\eta})$ differs from $\{0\}$ or $\emptyset$ only if $|P_0\hat{z}| = -\eta$. Consequently, the constitutive equations (7.3.9), (7.3.10) or (7.3.2) – (7.3.4) are of monotone-gradient type if and only if k satisfies (7.3.13).

If the constitutive equations are not of monotone-gradient type, we apply Theorem 7.2.1 to transform them. Conditions (i) und (ii) of this theorem are satisfied with the choice $M = \hat{L}$. The free energy in the transformed equations is then equal to the function ψ from (7.3.11). The transformation field H is determined by solving (7.2.5) with $g_K = g$. Since the set K defined in (7.3.12) is convex if a vector field H with the form (7.2.4) can be found which solves the system of partial differential equations (7.2.5), and for which the set $K^* = -\hat{L}H(-\hat{L}^{-1}K)$ is convex and satisfies $0 \in K^*$, then the assumptions of Theorem 7.2.1 will be satisfied and therefore transformation to monotone-gradient type will be possible

To determine the form which the partial differential equation (7.2.5) assumes in the present example, we first compute the subdifferential on the right hand side of (7.2.5). Observe that (7.3.12) yields

$$-\hat{L}^{-1}K = \left\{ z \in \mathcal{S}^3 \times \mathbb{R} \;\middle|\; -\hat{L}z = -\hat{L}(\hat{z},\eta) = -(D\hat{z},\eta) \in K \right\}$$

$$= \left\{ (\hat{z},\eta) \;\middle|\; |P_0 D\hat{z}| \leq \eta \right\} \subseteq \mathcal{S}^3 \times \mathbb{R}_0^+.$$

With this formula for $-\hat{L}^{-1}K$ we see exactly as in Lemma 7.3.1 that for $z = (\hat{z},\eta) \in \mathcal{S}^3 \times \mathbb{R}$ with $\eta \neq 0$ the subdifferential satisfies

$$\partial\chi_{-\hat{L}^{-1}K}(z) = \partial\chi\left(\frac{|P_0 D\hat{z}|}{\eta}\right)\left(D\hat{g}(D\hat{z}), -1\right). \tag{7.3.14}$$

To compute the term on the left hand side of (7.2.5) we note first that (7.3.6) and (7.3.7) imply for $z = (\hat{z},\eta) \in \mathcal{S}^3 \times \mathbb{R}^+$

$$g(-\hat{L}z) = g\left(-\hat{L}(\hat{z},\eta)\right) = g(-D\hat{z}, -\eta) \tag{7.3.15}$$

$$= \partial\chi\left(\frac{|P_0 D\hat{z}|}{\eta}\right)\left(-\hat{g}(D\hat{z}), k(|P_0 D\hat{z}|,\eta)\right).$$

The operator P in (7.2.4) is the projection along $\ker(B)$ onto the subspace $R(\hat{L}^{-1}B^T D)$. From (7.3.5) and (7.3.6) we deduce that $\hat{L}^{-1}B^T D\varepsilon = (\varepsilon, 0)$, hence $R(\hat{L}^{-1}B^T D) = \mathcal{S}^3 \times \{0\}$ and $\ker(B) = \{0\} \times \mathbb{R}$. Thus, P and Q are orthogonal projections with

$$Pz = P(\hat{z},\eta) = (\hat{z},0), \quad Qz = Q(\hat{z},\eta) = (0,\eta).$$

For a vector field of the form (7.2.4) we then have

$$H(z) = H(\hat{z},\eta) = Pz + \tilde{H}(Qz) = (\hat{z},0) + \tilde{H}(0,\eta) = (\hat{z}, \tilde{H}(\eta)). \tag{7.3.16}$$

To obtain the last equality we identified the mapping $\tilde{H} : \{0\} \times \mathbb{R} \to \{0\} \times \mathbb{R}$ with a vector field $\tilde{H} : \mathbb{R} \to \mathbb{R}$. Consequently, with $M = L$

$$H'(z) = H'(z)^T = \begin{pmatrix} I & 0 \\ 0 & \tilde{H}'(\eta) \end{pmatrix}, \quad H'(z)^T M H'(z) = \begin{pmatrix} D & 0 \\ 0 & \tilde{H}'(\eta)^2 \end{pmatrix}.$$

We combine this result with (7.3.15) to obtain for the left hand side of (7.2.5)

$$-H'(z)^T M H'(z)g(-\hat{L}z) = \partial\chi\left(\frac{|P_0 D\hat{z}|}{\eta}\right)\left(D\hat{g}(D\hat{z}), -\tilde{H}'(\eta)^2 k(|P_0 D\hat{z}|,\eta)\right).$$

Comparison with (7.3.14) shows that the partial differential equation (7.2.5) is satisfied for all $z = (\hat{z},\eta) \in \mathcal{S}^3 \times \mathbb{R}^+$ if and only if

$$\tilde{H}'(\eta)^2 k(\eta,\eta) = 1 \tag{7.3.17}$$

for all $\eta > 0$, since for $\eta \neq |P_0 D\hat{z}|$ the term $\partial\chi(\frac{|P_0 D\hat{z}|}{\eta})$ is equal to $\{0\}$ or to $\emptyset$. We can normalize $\tilde{H}$ to be an increasing function. With this normalization, from (7.3.17) we obtain the ordinary differential equation

$$\tilde{H}'(\eta) = \frac{1}{\sqrt{k(\eta, \eta)}}, \qquad (7.3.18)$$

which must be satisfied for all $\eta > 0$. As initial condition we choose

$$\hat{H}(0) = 0, \qquad (7.3.19)$$

which determines $\tilde{H}$ uniquely on the interval $[0, \infty)$. We continue $\tilde{H}$ to the interval $(-\infty, 0)$ as a smooth, strictly increasing function. Together with (7.3.16) this defines the transformation field H on all of $\mathcal{S}^3 \times \mathbb{R}$. By construction, H satisfies the partial differential equation (7.2.5) on $\mathcal{S}^3 \times \mathbb{R}^+$. This equation is satisfied on all of $\mathcal{S}^3 \times \mathbb{R}$ if we extend $g \circ (-\hat{L}^{-1})$ to $\mathcal{S}^3 \times \mathbb{R}_0^-$ by

$$g(-\hat{L}z) = [-H'(z)^T \hat{L} H'(z)]^{-1} \partial\chi_{-\hat{L}^{-1}K}(z).$$

It remains to examine whether the equations obtained by transformation of (7.3.9), (7.3.10) with this field are of monotone-gradient type. According to Theorem 7.2.1, the transformed equations are of this type if the set $K^* = -\hat{L}H(-\hat{L}^{-1}K)$ is convex and contains 0. Since (7.3.6) and (7.3.16) imply

$$\left[-\hat{L} \circ H \circ (-\hat{L}^{-1})\right]^{-1}(z, \eta) = -\hat{L}H^{-1}\left(-\hat{L}^{-1}(z, \eta)\right)$$

$$= -\hat{L}\left(-D^{-1}\hat{z}, \tilde{H}^{-1}(-\eta)\right) = \left(\hat{z}, -\tilde{H}^{-1}(-\eta)\right),$$

we obtain from (7.3.12)

$$K^* = -\hat{L}H(-\hat{L}^{-1}K) = \left\{z \,\middle|\, \left(\left[-\hat{L} \circ H \circ (-\hat{L}^{-1})\right]^{-1}z\right) \in K\right\}$$

$$= \left\{(\hat{z}, \eta) \,\middle|\, \left(\hat{z}, -\tilde{H}^{-1}(-\eta)\right) \in K\right\}$$

$$= \left\{(\hat{z}, \eta) \,\middle|\, |P_0\hat{z}| \leq \tilde{H}^{-1}(-\eta)\right\} \qquad (7.3.20)$$

$$= \left\{(\hat{z}, \eta) \,\middle|\, \tilde{H}\left(|P_0\hat{z}|\right) \leq -\eta\right\},$$

where in the last step we used that $\tilde{H}$ is strictly increasing. Since $\tilde{H}(0) = 0$, by the initial condition (7.3.19), it follows from (7.3.20) that

$$0 \in K^* \subseteq \mathcal{S}^3 \times \mathbb{R}_0^-,$$

and that K^* is convex if and only if $H''(\eta) \geq 0$ for all $\eta > 0$. Because (7.3.18) implies

$$\tilde{H}''(\eta) = -\frac{1}{2}\, k(\eta, \eta)^{-3/2} \frac{\partial}{\partial\eta} k(\eta, \eta),$$

we see that K^* is convex if and only if

$$\frac{\partial}{\partial\eta} k(\eta, \eta) \leq 0 \qquad (7.3.21)$$

for all $\eta > 0$.

In summary, we obtain that the constitutive equations (7.3.9), (7.3.10) and there-fore also (7.3.2) – (7.3.4) can be transformed to monotone-gradient type, if the function

$$(\zeta \mapsto k(\zeta, \zeta)) \in C^1((0, \infty), (0, \infty))$$

satisfies (7.3.21), or equivalently, is a decreasing function.

7.4 Example 2: Several Variables of Isotropic Hardening

To demonstrate that models containing parameters of kinematic hardening can also be transformed to monotone type, we next consider such a model. This model is also more general than the model in the preceding example in that it contains finitely many parameters $\zeta_1, \ldots, \zeta_m$ of isotropic hardening.

The system of equations we consider is

$$\rho u_{tt} = \operatorname{div}_x T \tag{7.4.1}$$

$$T = D\big(\varepsilon(\nabla_x u) - \varepsilon_p\big) \tag{7.4.2}$$

$$\frac{\partial}{\partial t}\varepsilon_p \in \partial \chi \left(\frac{\big|P_0\big(T - \mathcal{M}(\varepsilon_p - \varepsilon_n)\big)\big|}{\Gamma(\zeta)} \right) \frac{P_0\big(T - \mathcal{M}(\varepsilon_p - \varepsilon_n)\big)}{\big|P_0\big(T - \mathcal{M}(\varepsilon_p - \varepsilon_n)\big)\big|} \tag{7.4.3}$$

$$\frac{\partial}{\partial t}\varepsilon_n \in \partial \chi\big(|\mathcal{M}(\varepsilon_p - \varepsilon_n)|\big) \frac{\varepsilon_p - \varepsilon_n}{|\varepsilon_p - \varepsilon_n|} \tag{7.4.4}$$

$$\frac{\partial}{\partial t}\zeta = \big|\frac{\partial}{\partial t}\varepsilon_p\big| k\big(|P_0\big(T - \mathcal{M}(\varepsilon_p - \varepsilon_n)\big)|, \zeta\big). \tag{7.4.5}$$

Here $\mathcal{M}$ is a positive constant, and $\zeta = (\zeta_1, \ldots, \zeta_m) \in \mathbb{R}^m$ with $m \geq 1$ is the vector of parameters of isotropic hardening. The parameter $\varepsilon_n = \varepsilon_n(x, t) \in \mathcal{S}^3$ is interpreted as an inelastic strain tensor, and

$$\Gamma : D(\Gamma) \subseteq \mathbb{R}^m \to [0, \infty), \quad k : [0, \infty) \times D(\Gamma) \to \mathbb{R}^m$$

are given, continuously differentiable functions. Γ is assumed to be a concave func-tion with a closed, convex domain $D(\Gamma)$. This means that

$$\Gamma\big(r\zeta + (1 - r)\overline{\zeta}\big) \geq r\Gamma(\zeta) + (1 - r)\Gamma(\overline{\zeta})$$

holds for all $\zeta, \overline{\zeta} \in D(\Gamma)$ and $0 \leq r \leq 1$. For simplicity we also assume that $\Gamma(\zeta) = 0$ if and only if $\zeta \in \partial(D(\Gamma))$, hence $\{\zeta \in D(\Gamma) \mid \Gamma(\zeta) \neq 0\} = D(\Gamma)^0$, and that

$$0 \in D(\Gamma).$$

Moreover, it is assumed that k and Γ satisfy

$$k(\Gamma(\zeta), \zeta) \cdot \nabla\Gamma(\zeta) \geq 0 \tag{7.4.6}$$

for all $\zeta \in D(\Gamma)$. The other notations are as in the preceding example. The rheological diagram corresponding to this system of equations is displayed in figure 7.4.1. The friction element 1 in this diagram can harden.

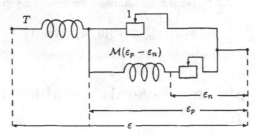

Figure 7.4.1

The initial conditions for the solutions of the system (7.4.1) – (7.4.5) are chosen such that $\Gamma(\zeta(x,0)) \geq \delta > 0$. Since (7.4.5) yields

$$\frac{\partial}{\partial t}\Gamma(\zeta) = \nabla\Gamma(\zeta) \cdot \frac{\partial}{\partial t}\zeta = |\frac{\partial}{\partial t}\varepsilon_p| \, \nabla\Gamma(\zeta) \cdot k(|P_0(T - M(\varepsilon_p - \varepsilon_n))|, \zeta),$$

and since (7.4.3) and the definition of χ imply that $|\frac{\partial}{\partial t}\varepsilon_p| \neq 0$ only for $|P_0(T - M(\varepsilon_p - \varepsilon_n))| = \Gamma(\zeta)$, it follows from (7.4.6) that $\frac{\partial}{\partial t}\Gamma(\zeta) \geq 0$. Hence $\Gamma(\zeta(x,t)) \geq \Gamma(\zeta(x,0)) \geq \delta$ for all $t \geq 0$. Condition (7.4.6) thus implies that softening of the friction element 1 is impossible. Therefore, since the values of $\Gamma(\zeta)$ are always greater than zero, in the following investigations we can redefine the constitutive equations (7.4.3) – (7.4.5) at points $(\varepsilon_p, \varepsilon_n, \eta)$ with $\eta \in \partial(D(\Gamma)) = \{\eta \mid \Gamma(\eta) = 0\}$ in a way which reduces the technicalities of the investigations.

To show that also these equations are of pre-monotone type we set $N = 12 + m$, choose $z = (\varepsilon_p, \varepsilon_n, \zeta) \in \mathcal{S}^3 \times \mathcal{S}^3 \times \mathbb{R}^m \cong \mathbb{R}^N$ for the vector of internal variables, and define the map $B : \mathcal{S}^3 \times \mathcal{S}^3 \times \mathbb{R}^m \to \mathcal{S}^3$ by

$$Bz = B(\hat{z}, \tilde{z}, \eta) = \hat{z} \in \mathcal{S}^3. \tag{7.4.7}$$

For the transpose $B^T : \mathcal{S}^3 \to \mathcal{S}^3 \times \mathcal{S}^3 \times \mathbb{R}^m$ we thus obtain $B^T\hat{z} = (\hat{z}, 0, 0)$, hence $B^T DB(\hat{z}, \tilde{z}, \eta) = (D\hat{z}, 0, 0)$ and $B^T D\varepsilon = (D\varepsilon, 0, 0)$. We define the symmetric map $\hat{L} : \mathcal{S}^3 \times \mathcal{S}^3 \times \mathbb{R}^m \to \mathcal{S}^3 \times \mathcal{S}^3 \times \mathbb{R}^m$ by

$$\hat{L}z = \hat{L}(\hat{z}, \tilde{z}, \eta) = (D\hat{z} + M(\hat{z} - \tilde{z}), -M(\hat{z} - \tilde{z}), \eta). \tag{7.4.8}$$

This mapping is positive definite and $L = \hat{L} - B^T DB$ is positive semi-definite, since

$$(\hat{L}z) \cdot z = (D\hat{z}) \cdot \hat{z} + M|\hat{z} - \tilde{z}|^2 + |\eta|^2 > 0$$

for $z = (\hat{z}, \tilde{z}, \eta) \neq 0$, and

$$(Lz) \cdot z = M|\hat{z} - \tilde{z}|^2 + |\eta|^2 \geq 0.$$

Finally, noting that by our assumption $\Gamma(\eta) \neq 0$ for $\eta \in D(\Gamma)^0$, we define the map

$$g : \mathcal{S}^3 \times \mathcal{S}^3 \times (-D(\Gamma)^0) \to \mathcal{P}(\mathcal{S}^3 \times \mathcal{S}^3 \times \mathbb{R}^m)$$

by

$$g(z) = g(\hat{z}, \tilde{z}, \eta) = \partial\chi\Big(\frac{|P_0\hat{z}|}{\Gamma(-\eta)}\Big)\big(\hat{g}(\hat{z}), 0, k(|P_0\hat{z}|, -\eta)\big) + \partial\chi(|\tilde{z}|)(0, \tilde{z}, 0), \quad (7.4.9)$$

where $\hat{g}(\hat{z}) = \frac{P_0\hat{z}}{|P_0\hat{z}|}$, and where we use the convention to define the sum of two sets Λ_1, Λ_2 by

$$\Lambda_1 + \Lambda_2 = \{\lambda_1 + \lambda_2 \mid \lambda_1 \in \Lambda_1, \ \lambda_2 \in \Lambda_2\}.$$

In particular, $\Lambda_1 + \Lambda_2 = \emptyset$ if $\Lambda_1 = \emptyset$ or $\Lambda_2 = \emptyset$. With these definitions, the equations (7.4.1) – (7.4.5) can be written in the form

$$\rho u_{tt} = \operatorname{div}_x T \tag{7.4.10}$$

$$T = D\big(\varepsilon - Bz\big) \tag{7.4.11}$$

$$z_t \in g(\overline{L}\varepsilon - \hat{L}z) = g(-\rho\nabla_z\psi(\varepsilon, z)) \tag{7.4.12}$$

with $\overline{L} = B^T D$ and with the free energy

$$\rho\psi(\varepsilon, z) = \frac{1}{2}[D(\varepsilon - Bz)] \cdot (\varepsilon - Bz) + \frac{1}{2}(Lz) \cdot z$$

$$= \frac{1}{2}[D(\varepsilon - \hat{z})] \cdot (\varepsilon - \hat{z}) + \frac{1}{2}\mathcal{M}|\hat{z} - \tilde{z}|^2 + \frac{1}{2}|\eta|^2, \quad z = (\hat{z}, \tilde{z}, \eta).$$

The equations (7.4.11), (7.4.12) satisfy the conditions of Definition 3.1.1. Therefore the constitutive equations (7.4.2) – (7.4.5) are of pre-monotone type.

$g(z)$ defined in (7.4.9) differs from $\{0\}$ and from $\emptyset$ only if z belongs to the boundary of the closed set

$$K = \Big\{z = (\hat{z}, \tilde{z}, \eta) \in \mathcal{S}^3 \times \mathcal{S}^3 \times \mathbb{R}^m \ \Big| \ -\eta \in D(\Gamma), \ |P_0\hat{z}| \le \Gamma(-\eta), \ |\tilde{z}| \le 1\Big\}. \tag{7.4.13}$$

This set contains 0, since we assumed that $0 \in D(\Gamma)$, hence $\Gamma(0) \ge 0$, and it is convex, since we assumed that Γ is concave. Therefore the constitutive equations (7.4.2) – (7.4.5) are of monotone-gradient type if and only if we can extend g to all of $\mathcal{S}^3 \times \mathcal{S}^3 \times \mathbb{R}^m$ such that the extension is equal to the subdifferential $\partial\chi_K$ of the characteristic function χ_K of K. It is not difficult to show that for $z = (\hat{z}, \tilde{z}, \eta) \in K$ with $-\eta \in D(\Gamma)^0$, hence $\Gamma(-\eta) \ne 0$, this subdifferential satisfies

$$\partial\chi_K(z) = \partial\chi\Big(\frac{|P_0\hat{z}|}{\Gamma(-\eta)}\Big)\big(\hat{g}(\hat{z}), 0, \nabla\Gamma(-\eta)\big) + \partial\chi(|\tilde{z}|)\,(0, \tilde{z}, 0). \tag{7.4.14}$$

Comparison with (7.4.9) shows that if we extend g to the set of all $z \in \mathcal{S}^3 \times \mathcal{S}^3 \times (\mathbb{R}^m \backslash (-D(\Gamma)^0))$ by $g(z) = \partial\chi_K(z)$, then g is equal to $\partial\chi_K$ on all of $\mathcal{S}^3 \times \mathcal{S}^3 \times \mathbb{R}^m$, if and only if

$$k(\Gamma(\eta), \eta) = \nabla\Gamma(\eta)$$

for all $\eta \in D(\Gamma)$. Consequently, the constitutive equations (7.4.2) –(7.4.5) are of monotone-gradient type if and only if this equality holds. This generalizes condition (7.3.13) of example 1, where we have $\Gamma(\eta) = \eta$. Hence $\Gamma'(\eta) = 1$.

If the constitutive equations are not of monotone-gradient type, we apply Theorem 7.2.1 to transform them. The conditions (i) and (ii) of this theorem are satisfied if we choose $M = \hat{L}$. The transformation field H in this theorem is assumed to be of the form (7.2.4) and must satisfy the partial differential equation (7.2.5). For H we make the ansatz

$$H(z) = H(\hat{z}, \tilde{z}, \eta) = (\hat{z}, \tilde{z}, \overline{H}(\eta)) \tag{7.4.15}$$

for a vector field $\overline{H} : \mathbb{R}^m \to \mathbb{R}^m$. To show that such transformation fields are of the form (7.2.4), we determine the projections P and Q in (7.2.4). To this end we note that (7.4.8) yields

$$\hat{L}^{-1}(\hat{z}, \tilde{z}, \eta) = \left(D^{-1}(\hat{z} + \tilde{z}), D^{-1}(\hat{z} + \tilde{z}) + \frac{1}{M}\tilde{z}, \eta\right),$$

hence, together with the definition of B in (7.4.7),

$$\hat{L}^{-1}\overline{L}\varepsilon = \hat{L}^{-1}B^T D\varepsilon = \hat{L}^{-1}(D\varepsilon, 0, 0) = (\varepsilon, \varepsilon, 0).$$

This yields

$$R(\hat{L}^{-1}\overline{L}) = \{(\varepsilon, \varepsilon, 0) \mid \varepsilon \in \mathcal{S}^3\} \subseteq \mathcal{S}^3 \times \mathcal{S}^3 \times \mathbb{R}^m$$
$$\ker(B) = \{0\} \times \mathcal{S}^3 \times \mathbb{R}^m,$$

whence, since P is the projection onto $R(\hat{L}^{-1}\overline{L})$ along $\ker(B)$,

$$P(\hat{z}, \tilde{z}, \eta) = (\hat{z}, \hat{z}, 0), \quad Q(\hat{z}, \tilde{z}, \eta) = (I - P)(\hat{z}, \tilde{z}, \eta) = (0, \tilde{z} - \hat{z}, \eta).$$

From these equations it follows that the vector field H defined in (7.4.15) is of the form (7.2.4), since if we define $\tilde{H} : \ker(B) \to \ker(B)$ by $\tilde{H}(0, \tilde{z}, \eta) = (0, \tilde{z}, \overline{H}(\eta))$, then

$$H(z) = H(\hat{z}, \tilde{z}, \eta) = (\hat{z}, \tilde{z}, \overline{H}(\eta)) = (\hat{z}, \hat{z}, 0) + (0, \tilde{z} - \hat{z}, \overline{H}(\eta))$$

$$= Pz + \tilde{H}(0, \tilde{z} - \hat{z}, \eta) = Pz + \tilde{H}(Qz),$$

which is (7.2.4).

Next we compute the form which the partial differential equation (7.2.5) assumes for such vector fields. Since

$$H'(z) = \begin{pmatrix} I & 0 \\ 0 & \overline{H}'(\eta) \end{pmatrix}, \quad H'(z)^T = \begin{pmatrix} I & 0 \\ 0 & \overline{H}'(\eta)^T \end{pmatrix}, \tag{7.4.16}$$

where I is the identity on $\mathcal{S}^3 \times \mathcal{S}^3$, we obtain with $M = \hat{L}$ from (7.4.8) that

$$H'(z)^T M H'(z) = H'(z)^T \hat{L} H'(z) = \hat{L} H'(z)^T H'(z).$$

Equation (7.2.5) can therefore be written as

$$- H'(z)^T H'(z)\, g(-\hat{L}z) = \hat{L}^{-1}\partial\chi_{-\hat{L}^{-1}K}(z). \tag{7.4.17}$$

With the notation $\hat{L}(\hat{z}, \tilde{z}, \eta) = (\hat{w}, \tilde{w}, \eta)$ we obtain from (7.4.9) and (7.4.16) for the left hand side of this equation

$$-H'(z)^T H'(z) g(-\hat{L}z) = -H'(z)^T H'(z) g(-\hat{w}, -\tilde{w}, -\eta)$$

$$= -H'(z)^T H'(z) \Big[\partial\chi\Big(\frac{|P_0\hat{w}|}{\Gamma(\eta)}\Big) \big(-\hat{g}(\hat{w}), 0, k(|P_0\hat{w}|, \eta) \big)$$

$$+ \partial\chi(|\tilde{w}|)(0, -\tilde{w}, 0) \Big] \qquad (7.4.18)$$

$$= \partial\chi\Big(\frac{|P_0\hat{w}|}{\Gamma(\eta)}\Big) \big(\hat{g}(\hat{w}), 0, -\overline{H}'(\eta)^T \overline{H}'(\eta) k(|P_0\hat{w}|, \eta) \big)$$

$$+ \partial\chi(|\tilde{w}|)(0, \tilde{w}, 0).$$

To compute the right hand side of (7.4.17) we use Lemma 7.1.2. If for H in this lemma we set the linear symmetric mapping $-\hat{L}^{-1}$, then we obtain

$$\partial\chi_K(z) = -\hat{L}^{-1} \partial\chi_{-\hat{L}^{-1}K}(-\hat{L}^{-1}z),$$

hence

$$\hat{L}^{-1} \partial\chi_{-\hat{L}^{-1}K}(z) = -\partial\chi_K(-\hat{L}z).$$

We insert (7.4.14) into this equation and again use the notation $\hat{L}(\hat{z}, \tilde{z}, \eta) = (\hat{w}, \tilde{w}, \eta)$ to obtain

$$\hat{L}^{-1} \partial\chi_{-\hat{L}^{-1}K}(z) = -\partial\chi_K(-\hat{w}, -\tilde{w}, -\eta)$$

$$= \partial\chi\Big(\frac{|P_0\hat{w}|}{\Gamma(\eta)}\Big) \big(\hat{g}(\hat{w}), 0, -\nabla\Gamma(\eta) \big) + \partial\chi(|\tilde{w}|)(0, \tilde{w}, 0).$$

Comparison with (7.4.18) shows that (7.4.17) and therefore the partial differential equation (7.2.5) is satisfied if and only if the system of m partial differential equations

$$\overline{H}'(\eta)^T \overline{H}'(\eta) \, k(\Gamma(\eta), \eta) = \nabla\Gamma(\eta) \qquad (7.4.19)$$

holds for all $\eta \in D(\Gamma)$, since $\partial\chi(\frac{|P_0\hat{w}|}{\Gamma(\eta)})$ is equal to $\{0\}$ or $\emptyset$ if $|P_0\hat{w}| \neq \Gamma(\eta)$.

Thus, if a solution $\overline{H}$ of (7.4.19) exists, then, according to Theorem 7.2.1, the vector field H defined in (7.4.15) transforms the constitutive equations (7.4.11), (7.4.12) and therefore (7.4.2) – (7.4.5) to monotone-gradient type, if the set $K^* = -\hat{L}H(-\hat{L}^{-1}K)$ is convex and contains 0. To derive sufficient conditions which guarantee that K^* has these properties, note that the definitions of $\hat{L}$ and H in (7.4.8) and (7.4.15) imply $-\hat{L}H^{-1}(-\hat{L}^{-1}(\hat{z}, \tilde{z}, \eta)) = (\hat{z}, \tilde{z}, -\overline{H}^{-1}(-\eta))$. From the definition of K in (7.4.13) we thus deduce that

$$K^* = \{z \mid -\hat{L}H^{-1}(-\hat{L}^{-1}z) \in K\}$$

$$= \{(\hat{z}, \tilde{z}, -\eta) \mid \eta \in \overline{H}(D(\Gamma)), |P_0\hat{z}| \leq \Gamma(\overline{H}^{-1}(\eta)), |\tilde{z}| \leq 1\}.$$

Since we assumed that $0 \in D(\Gamma)$, we see from this formula that if the vector field $\overline{H} : D(\Gamma) \to \mathbb{R}^m$ transforms $D(\Gamma)$ onto itself, then $0 \in K^*$. This formula also shows that K^* is convex if the function $\Gamma \circ \overline{H}^{-1}$ is concave. This is satisfied if the Hessian matrix $\nabla^2(\Gamma \circ \overline{H}^{-1})(\eta)$ is negative semi-definite for all $\eta \in \overline{H}(D(\Gamma)) = D(\Gamma)$.

In summary, if a solution $\overline{H}$ of the system of partial differential equations (7.4.19) exists, which transforms $D(\Gamma)$ onto itself, and if for this solution the Hessian matrix $\nabla^2(\Gamma \circ \overline{H}^{-1})$ is negative semi-definite on $D(\Gamma)$, then H defined in (7.4.15) transforms the constitutive equations (7.4.2) – (7.4.5) to monotone-gradient type.

To find such a solution $\overline{H}$, one can prescribe boundary conditions (or initial conditions) which are convenient for the particular problem considered. For example,

$$\overline{H}(\eta) = \eta, \quad \eta \in \partial(D(\Gamma)),$$

could be a useful condition, since it guarantees that the boundary $\partial(D(\Gamma))$ is mapped onto itself. The partial differential equation (7.4.19) together with such a condition forms a boundary value problem for $\overline{H}$.

This completes our discussion of example 2. In the same way one can study and transform a more complicated constitutive model, for which both friction elements in figure 7.4.1 can harden, rather than only the friction element 1. It seems however, that the transformation theory developed in this work cannot be applied to constitutive models where the hardening behavior of one friction element is influenced by the deformation behavior of the second friction element, since the resulting constitutive equations do not satisfy the classificaton conditions stated in Theorem 6.2.4, and thus fall outside the class $\mathcal{TG}$. Rate dependent constitutive models having this property are, for example, the models of Chaboche (Example 3 in Section 2.2), and of Méric, Poubanne and Cailletaud (Example 9 in Section 2.2). To study such models, a more general transformation theory with transformation fields depending on ε must be developed.

Chapter 8

Application of the Theory to Engineering Models

As an example for the transformation theory for rate dependent constitutive equations developed in Chapters 5 and 6, in this chapter we apply the theory to the constitutive model of Bodner and Partom introduced in Section 2.2. We choose this model not because of its usefulness in solid mechanics and engineering, which we are not able to judge, but because on the one hand, mathematically it is relatively simple, and on the other hand, the results obtained for this model concerning the applicability of the transformation theory are respresentative for other models. In the following investigations the limitations of the transformation theory also become obvious. As is discussed more precisely in Chapter 9, these limitations are mainly caused by the fact that the transformation fields used in this book are not allowed to depend on ε.

It is noteworthy that in example 7 of the following Section 8.5, a constitutive equation is given which belongs to both $T\mathcal{G}$ and $T\mathcal{M}$, but not to $T(\mathcal{M}\mathcal{G})$. Therefore this example shows that

$$T(\mathcal{M}\mathcal{G}) \subset T\mathcal{M} \cap T\mathcal{G}$$

with equality excluded.

8.1 Pre-Monotone Type of the Model

For the model of Bodner and Partom introduced in example 4 of Section 2.2 the equations $(2.1.1) - (2.1.3)$ have the form

$$\rho u_{tt} = \operatorname{div}_x T \tag{8.1.1}$$

$$T = D\big(\varepsilon(\nabla_x u) - \varepsilon_p\big) \tag{8.1.2}$$

$$\frac{\partial}{\partial t}\,\varepsilon_p = F\Big(\frac{|P_0 T|}{\zeta}\Big)\,\frac{P_0 T}{|P_0 T|} \tag{8.1.3}$$

$$\frac{\partial}{\partial t}\,\zeta = \gamma(\zeta)\,F\Big(\frac{|P_0 T|}{\zeta}\Big)\,|P_0 T| - \delta_A(\zeta), \tag{8.1.4}$$

where $P_0 : \mathcal{S}^3 \to \mathcal{S}^3$ is the orthogonal projector to the subspace $\{\sigma \in \mathcal{S}^3 \mid \operatorname{trace}(\sigma) = 0\}$ and $P_0 T(x,t)$ is thus the stress deviator, $\varepsilon_p(x,t) \in \mathcal{S}^3$ is the plastic strain tensor,

$\zeta(x,t) > 0$ is the hardening parameter, and

$$F : \mathbb{R}_0^+ \to \mathbb{R}_0^+, \quad \gamma : D(\gamma) \subseteq \mathbb{R}^+ \to \mathbb{R}^+, \quad \delta_A : D(\delta_A) \subseteq \mathbb{R}^+ \to \mathbb{R}_0^+ \qquad (8.1.5)$$

are given functions. We call F, γ and δ material functions. In the model they are chosen to be $F = F_1$, $\gamma = \gamma_1$ with

$$F_1(s) = d \exp\left(-\alpha\left(\frac{1}{s}\right)^n\right), \quad \alpha = \frac{n/2 + 1}{n} \qquad (8.1.6)$$

$$\gamma_1(\zeta) = m(\zeta_1 - \zeta), \quad D(\gamma_1) = [\zeta_2, \zeta_1) \qquad (8.1.7)$$

and

$$\delta_A(\zeta) = A\zeta_1\left(\frac{\zeta - \zeta_2}{\zeta_1}\right)^r, \quad D(\delta_A) = [\zeta_2, \infty). \qquad (8.1.8)$$

Here $n, r > 1$; $d, m > 0$; $A \geq 0$; $\zeta_1 > \zeta_2 > 0$ are constants depending on the material considered. To test the applicability of the transformation theory we also study the model with a second choice for the material functions, namely $F = F_2$ and $\gamma = \gamma_2$ with

$$F_2(s) = ds^n \qquad (8.1.9)$$

$$\gamma_2(\zeta) = m\zeta^{-r}, \quad D(\gamma_2) = [\zeta_2, \infty). \qquad (8.1.10)$$

d, m, r, $\zeta_2 > 0$, $n > 1$ are constants. In fact, as noted in Section 2.2, the power law (8.1.9) is often used in constitutive models, whereas the function F_1 from (8.1.6) is a rather special choice.

We first show that the constitutive equations (8.1.2) – (8.1.4) are of pre-monotone type. This can be shown in a similar way as for the model of Chaboche in Section 3.3. As in the investigation of that model and of the other models in Chapter 3 the isomorphism defined in (3.3.1) is used to identify $\mathcal{S}^3$ with $\mathbb{R}^6$. We set $N = 7$ and choose $z = (\varepsilon_p, \zeta) \in \mathcal{S}^3 \times \mathbb{R} \cong \mathbb{R}^N$ for the vector of internal variables. The mapping $B : \mathcal{S}^3 \times \mathbb{R} \to \mathcal{S}^3$ is defined by

$$B(\sigma, \eta) = \sigma \in \mathcal{S}^3 \qquad (8.1.11)$$

with the transpose $B^T : \mathcal{S}^3 \to \mathcal{S}^3 \times \mathbb{R}$ given by $B^T \sigma = (\sigma, 0)$. We set

$$\overline{L} = B^T D : \mathcal{S}^3 \to \mathcal{S}^3 \times \mathbb{R}, \qquad (8.1.12)$$

hence $\overline{L}\varepsilon = (D\varepsilon, 0)$ and $B^T DBz = B^T DB(\sigma, \eta) = (D\sigma, 0)$. We further define the linear mapping $\hat{L} : \mathcal{S}^3 \times \mathbb{R} \to \mathcal{S}^3 \times \mathbb{R}$ by

$$\hat{L}z = \hat{L}(\sigma, \eta) = (D\sigma, \eta). \qquad (8.1.13)$$

Then $\hat{L}$ is symmetric, positive definite, and $L = \hat{L} - B^T DB$ is positive semi-definite because

$$(\hat{L}z) \cdot z = [\hat{L}(\sigma, \eta)] \cdot (\sigma, \eta) = (D\sigma) \cdot \sigma + \eta^2 > 0$$

for $z \neq 0$ and

$$(Lz) \cdot z = \eta^2 \geq 0.$$

If we finally define $g : -\left[S^3 \times (D(\gamma) \cap D(\delta_A))\right] \to S^3 \times \mathbb{R}_0^+$ by

$$g(\sigma, \eta) = \begin{pmatrix} g_1(\sigma, \eta) \\ g_2(\sigma, \eta) \end{pmatrix} = \begin{pmatrix} F(\frac{|\tau|}{-\eta}) \frac{\tau}{|\tau|} \\ \gamma(-\eta)F(\frac{|\tau|}{-\eta})|\tau| - \delta_A(-\eta) \end{pmatrix}, \tag{8.1.14}$$

where $\tau = P_0 \sigma$, then it is seen immediately that equations (8.1.1) – (8.1.4) can be written in the form

$$\rho u_{tt} = \operatorname{div}_x T \tag{8.1.15}$$

$$T = D(\varepsilon - Bz) \tag{8.1.16}$$

$$z_t = f(\varepsilon, z) = g(\overline{L}\varepsilon - \hat{L}z) = g(-\rho \nabla_z \psi(\varepsilon, z)) \tag{8.1.17}$$

with

$$\rho\psi(\varepsilon, z) = \frac{1}{2}[D(\varepsilon - Bz)] \cdot (\varepsilon - Bz) + \frac{1}{2}(Lz) \cdot z$$

$$= \frac{1}{2}[D(\varepsilon - \varepsilon_p)] \cdot (\varepsilon - \varepsilon_p) + \frac{1}{2}\zeta^2.$$

Here we set $z = (\varepsilon_p, \zeta)$ to obtain the last equality. Comparison with Definition 3.1.1 shows that the constitutive equations (8.1.16), (8.1.17) and consequently also (8.1.2) – (8.1.4) are of pre-monotone type.

8.2 Conditions for Monotone Type of the Model

We now use Lemma 3.4.1 to study whether the constitutive equations (8.1.16), (8.1.17) are of monotone type.

Proposition 8.2.1 (I) *Assume that* $F(0) = F'(0) = 0$, $F'(s) > 0$ *for* $s > 0$ *and that the domain* $D(\gamma) \cap D(\delta_A)$ *is an interval. Then the constitutive equations* (8.1.16), (8.1.17) *are of monotone type if and only if a constant* $a > 0$ *exists such that the two inequalities*

$$\frac{4}{a}\frac{F'}{\eta}\left(F'\frac{|\tau|^2}{\eta^2}\gamma - F|\tau|\gamma' + \delta_A'\right) \geq \left[\frac{1}{a}F'\frac{|\tau|}{\eta^2} + \left(F'\frac{|\tau|}{\eta} + F\right)\gamma\right]^2 \tag{8.2.1}$$

$$a\eta\gamma(\eta) \leq 1 \tag{8.2.2}$$

hold for all $(\tau, \eta) = (P_0\sigma, \eta)$ *with* $(\sigma, \eta) \in -D(g) = S^3 \times (D(\gamma) \cap D(\delta_A))$. *In* (8.2.1) *the argument of* F *and* F' *is* $|\tau|/\eta$, *and the argument of* γ, γ' *and* δ_A' *is* η.
(II) *If* $F = F_2$ *and* $A = 0$, *hence* $\delta_A = 0$, *then* (8.2.1) *is equivalent to*

$$\eta\frac{d}{d\eta}[\eta\gamma(\eta)] \leq -\frac{a}{4n}\left((n+1)\eta\gamma(\eta) - \frac{n}{a}\right)^2 \tag{8.2.3}$$

for all $\eta \in D(\gamma)$.

Example 1. If $F = F_1 = d\exp(-\alpha s^{-n})$ with $\alpha, d > 0, n > 1$, then the constitutive equations (8.1.16), (8.1.17) are not of monotone type, since (8.2.1) is not satisfied. To show this we use that such an F satisfies

$$F'(s) = \frac{\alpha n}{s^{n+1}} F(s).$$

If we insert this equation into (8.2.1) and set $s = |\tau|/\eta$, we obtain

$$\frac{4}{a} F(s)^2 \Big[\Big(\frac{\alpha n}{s^n}\Big)^2 \frac{\gamma(\eta)}{\eta} - \frac{\alpha n}{s^n} \gamma'(\eta) \Big] + \frac{4}{a} F(s) \frac{\alpha n}{s^{n+1}} \frac{\delta'_A(\eta)}{\eta}$$

$$\geq F(s)^2 \Big[\frac{\alpha n}{as^n \eta} + \Big(\frac{\alpha n}{s^n} + 1\Big) \gamma(\eta) \Big]^2,$$

which must be satisfied for all $s \geq 0$ and $\eta \in D(\gamma) \cap D(\delta_A)$. However, this inequality does not hold for all these values of (s, η), since $F(s) = d\exp(-\alpha s^{-n}) \to d$ for $s \to \infty$ implies that the left hand side of this inequality tends to zero for $s \to \infty$, whereas the right hand side tends to $d^2 \gamma(\eta)^2 > 0$. Therefore a constant $a > 0$ such that (8.2.1) holds for all values of $(\tau, \eta) \in S^3 \times (D(\gamma) \cap D(\delta_A))$ does not exist.

Example 2. If $F = F_2$, $\gamma = \gamma_2$ and $\delta_A = 0$, then the constitutive equations (8.1.16), (8.1.17) are of monotone type if and only if $r = 1$, because the function $\gamma(\eta) = \gamma_2(\eta) = m\eta^{-r}$ satisfies the inequalities (8.2.3) and (8.2.2) only if $r = 1$ and

$$a = \frac{n}{m(n+1)}.$$

To see this, note that with $r = 1$ the left hand side of (8.2.3) is equal to zero. Therefore the right hand side must also vanish, which happens only if $a = n/[m(n+1)]$. With this choice of a (8.2.2) is also satisfied, because $a\eta\gamma(\eta) = am = n/(n+1) < 1$. If $r < 1$ in (8.1.10), then the left hand side of (8.2.3) is positive, and the right hand side less than or equal to zero, whereas for $r > 1$ the left hand side of (8.2.3) satisfies

$$\eta \frac{d}{d\eta}[\eta\gamma(\eta)] = m(1-r)\eta^{-r+1} \to 0, \quad \eta \to \infty,$$

and the right hand side satisfies

$$-\frac{a}{4n}\Big((n+1)\eta\gamma(\eta) - \frac{n}{a}\Big)^2 \to -\frac{n}{4a} < 0, \quad \eta \to \infty.$$

Example 3. Let $F = F_2$, $\gamma(\eta) = \gamma_1(\eta) = m(\zeta_1 - \eta)$ with $D(\gamma_1) = [\zeta_2, \zeta_1]$ and $\delta_A = 0$. Then the following results hold. For $\zeta_2 < \zeta_1/2$ the constitutive equations (8.1.16), (8.1.17) are not of monotone type, since a constant a such that (8.2.3) holds for all $\eta \in [\zeta_2, \zeta_1]$ does not exist. For $\zeta_2 = \zeta_1/2$ the constitutive equations (8.1.16), (8.1.17) are of monotone type if and only if $n \leq 15$, because in this case (8.2.3) and (8.2.2) are satisfied for all $\eta \in [\zeta_2, \zeta_1)$ with the choice

$$a = \frac{4n}{m(n+1)\zeta_1^2}.$$

To see this, observe that for $\gamma = \gamma_1$ the left hand side of (8.2.3) is a quadratic polynomial with zeros in 0 and $\zeta_1/2$. On the interval $0 < \eta < \zeta_1/2$ this polynomial is positive, and therefore (8.2.3) cannot be satisfied on this interval. The right hand side of (8.2.3) is a non-positive fourth order polynomial. (8.2.3) can only hold on the whole interval $[\zeta_1/2, \zeta_1)$ if this polynomial has a zero at $\eta = \zeta_1/2$. This implies that a must have the value given above. With this value of a it is easily seen that (8.2.3) is satisfied on the interval $[\zeta_1/2, \zeta_1)$ if and only if $n \leq 15$. The inequality (8.2.2) holds on $[\zeta_1/2, \zeta_1)$ for this value of a since

$$a\eta\gamma(\eta) \leq a\gamma(\zeta_1/2)\zeta_1/2 = \frac{n}{n+1} < 1.$$

To prove Proposition 8.2.1, we need two lemmas:

Lemma 8.2.2 *If $F'(s) > 0$ for $s > 0$, then the kernel of the Jacobi-matrix of the function g defined in (8.1.14) satisfies*

$$\ker[B\nabla g(\sigma, \eta)] = \{\lambda(P_0\sigma, \eta) \mid \lambda \in \mathbb{R}\} \qquad (8.2.4)$$

for all $(\sigma, \eta) \in D(g)$ with $P_0\sigma \neq 0$, hence

$$\bigcap_{\varepsilon \in D(f,z)} \ker[B\nabla g(\overline{L}\varepsilon - \hat{L}z)] = \{0\} \qquad (8.2.5)$$

for all $z \in S^3 \times \mathbb{R}$ with $D(f, z) \neq \emptyset$, where $D(f, z) = \{\varepsilon \in S^3 \mid (\varepsilon, z) \in D(f)\}$, and

$$\bigcap_{(\varepsilon, z) \in D(f)} \ker[\nabla g(\overline{L}\varepsilon - \hat{L}z)] = \{0\}. \qquad (8.2.6)$$

Proof. For $\sigma \in S^3$ we set $\tau = P_0\sigma$ and identify $\tau \in S^3$ with the vector in $\mathbb{R}^9$ consisting of the components of the matrix τ. Using that $P_0 : S^3 \to S^3$ is an orthogonal projection, hence $P_0^T = P_0$, we obtain from (8.1.14) after a short computation

$$
\nabla g(\sigma, \eta) = \begin{pmatrix} \nabla_\sigma g_1 & \frac{\partial}{\partial \eta} g_1 \\ \nabla_\sigma g_2 & \frac{\partial}{\partial \eta} g_2 \end{pmatrix}
$$

$$
= \begin{pmatrix} \frac{F'}{-\eta} \frac{\tau\tau^T}{|\tau|^2} + \frac{F}{|\tau|}[I - \frac{\tau\tau^T}{|\tau|^2}] & F' \frac{|\tau|}{\eta^2} \frac{\tau}{|\tau|} \\ \gamma[F'\frac{|\tau|}{-\eta} + F]\frac{\tau^T}{|\tau|} & \gamma F' \frac{|\tau|^2}{\eta^2} - \gamma' F|\tau| + \delta'_A \end{pmatrix}, \qquad (8.2.7)
$$

and therefore, by definition of B in (8.1.11),

$$
B\nabla g(\sigma, \eta) = \left(\frac{F'}{-\eta} \frac{\tau\tau^T}{|\tau|^2} + \frac{F}{|\tau|}\left[I - \frac{\tau\tau^T}{|\tau|^2}\right] \quad F'\frac{|\tau|}{\eta^2} \frac{\tau}{|\tau|} \right). \qquad (8.2.8)
$$

Here τ^T denotes the transpose of the vector $\tau \in \mathbb{R}^9$ and the arguments of F, γ, δ_A are as in (8.1.14). The mapping

$$
w \to \left[I - \frac{\tau\tau^T}{|\tau|^2}\right]w = w - \frac{\tau \cdot w}{|\tau|^2}\tau : \mathbb{R}^9 \to \mathbb{R}^9
$$

is the orthogonal projection onto the orthogonal space of the vector $\tau = P_0\sigma$. Using this, (8.2.4) is obtained from (8.2.8).

Equation (8.2.5) is a consequence of (8.2.4). To see this, note that the definition of $\overline{L}$ and $\hat{L}$ in (8.1.12) and (8.1.13) yield for $z = (\sigma, \eta) \in \mathcal{S}^3 \times \mathbb{R}$ that

$$\overline{L}\varepsilon - \hat{L}(\sigma, \eta) = (D(\varepsilon - \sigma), -\eta). \qquad (8.2.9)$$

Since $D(f, (\sigma, \eta))$ consists of all $\varepsilon \in \mathcal{S}^3$ satisfying

$$\overline{L}\varepsilon - \hat{L}(\sigma, \eta) = (D(\varepsilon - \sigma), -\eta) \in D(g) = -\big[\mathcal{S}^3 \times (D(\gamma) \cap D(\delta_A))\big],$$

we conclude that if $D(f, (\sigma, \eta))$ is not empty, then $\eta \in (D(\gamma) \cap D(\delta_A))$ and $D(f, (\sigma, \eta)) = \mathcal{S}^3$. Using (8.2.4) and (8.2.9) we thus deduce (8.2.5) from

$$\bigcap_{\varepsilon \in D(f, (\sigma, \eta))} \ker[B\nabla g(\overline{L}\varepsilon - \hat{L}(\sigma, \eta))] = \bigcap_{\varepsilon \in \mathcal{S}^3} \{\lambda(P_0 D(\varepsilon - \sigma), -\eta) \mid \lambda \in \mathbb{R}\} = \{0\}.$$

Finally, (8.2.6) is a consequence of (8.2.5), since the nullspace of $\nabla g(\overline{L}\varepsilon - \hat{L}z)$ is a subset of the nullspace of $B\nabla g(\overline{L}\varepsilon - \hat{L}z)$. This proves the lemma.

Lemma 8.2.3 *If $F'(s) > 0$ for $s > 0$, then a symmetric, positive definite linear mapping $M : \mathcal{S}^3 \times \mathbb{R} \to \mathcal{S}^3 \times \mathbb{R}$ satisfies condition (ii) of Lemma 3.4.1 for the function f defined in (8.1.17), if and only if it has the form*

$$M(\tau, \zeta) = (D\tau, a\zeta), \quad (\tau, \zeta) \in \mathcal{S}^3 \times \mathbb{R}, \qquad (8.2.10)$$

for a positive constant a.

Proof. Using (8.1.12), and

$$f(\varepsilon, z) = g(\overline{L}\varepsilon - \hat{L}z),$$

condition (ii) of Lemma 3.4.1 takes the form

$$[\nabla g(\overline{L}\varepsilon - \hat{L}z)]\,[\hat{L}M^{-1} - I]\,\overline{L} = 0$$

for all $(\varepsilon, z) \in D(f)$. From (8.2.6) we deduce that this is equivalent to

$$[\hat{L}M^{-1} - I]\,\overline{L} = 0.$$

If M^{-1} is represented by $M^{-1}(\sigma, \eta) = (M_{11}\sigma + M_{12}\eta, M_{21}\sigma + a^{-1}\eta)$ with suitable linear mappings M_{11}, M_{12}, M_{21} and $a \in \mathbb{R}$, then, noting the definitions of $\overline{L}$ and $\hat{L}$ in (8.1.12) and (8.1.13), we see that equivalent to the last equation is

$$(DM_{11}\sigma, M_{21}\sigma) = (\sigma, 0)$$

for all $\sigma \in \mathcal{S}^3$. This relation is equivalent to $M_{11} = D^{-1}$ and $M_{21} = 0$. The matrix M is symmetric if and only if $M_{12} = M_{21}^T = 0$, and M is positive definite if and only if $a > 0$. This proves the lemma.

Proof of Proposition 8.2.1. The proof of (I) is based on Lemma 3.4.1. We first observe that

$$D_0(f) = -\hat{L}^{-1}D(g) = -\hat{L}^{-1}\big(-[\mathcal{S}^3 \times (D(\gamma) \cap D(\delta_A))]\big) = \mathcal{S}^3 \times (D(\gamma) \cap D(\delta_A))$$

is convex, since by assumption $D(\gamma) \cap D(\delta_A)$ is an interval, and that for a mapping M of the form (8.2.10) the equation $\overline{L}\varepsilon - \hat{L}z = 0$ holds on the subspace

$$V_M = \{(\varepsilon, z) \in \mathcal{S}^3 \times \mathcal{S}^3 \times \mathbb{R} \mid z = M^{-1}B^T D\varepsilon = \hat{L}^{-1}\overline{L}\varepsilon\}.$$

This implies together with (8.1.17) that $(\varepsilon_0, z_0) + V_M \subseteq D(f)$ for all $(\varepsilon_0, z_0) \in D(f)$. Consequently, for every M of the form (8.2.10) the constitutive equations (8.1.16), (8.1.17) satisfy the conditions (v) and (vi) of Lemma 3.4.1. To prove statement (I) of the proposition, it thus suffices to verify that the inequalities (8.2.1), (8.2.2) are equivalent to conditions (i) – (iv) of Lemma 3.4.1.

Since by assumption $F'(s) > 0$ for $s > 0$, we conclude from Lemma 8.2.3 that a mapping M satisfies condition (ii) of Lemma 3.4.1 if and only if it is of the form (8.2.10). We show that for mappings of this form, condition (iii) of Lemma 3.4.1 is equivalent to (8.2.1).

To prove this, note that condition (iii) means that the matrix

$$-\nabla_z f(0, z) M^{-1} = -\nabla_z g(-\hat{L}z) M^{-1} = [(\nabla g)(-\hat{L}z)] \hat{L}M^{-1}$$

is positive semi-definite for all $z \in D_0(f)$. Since

$$\hat{L}M^{-1} = \begin{pmatrix} D & 0 \\ 0 & 1 \end{pmatrix} \begin{pmatrix} D^{-1} & 0 \\ 0 & \frac{1}{a} \end{pmatrix} = \begin{pmatrix} I & 0 \\ 0 & \frac{1}{a} \end{pmatrix}$$

and since $\hat{L}$ is invertible, an equivalent condition is that

$$J(\sigma, \eta) = \nabla g(\sigma, \eta) \begin{pmatrix} I & 0 \\ 0 & \frac{1}{a} \end{pmatrix}$$

is positive semi-definite for all $(\sigma, \eta) \in D(g)$. Since we assumed that $F(0) = F'(0) = 0$, it follows from (8.2.7) that ∇g is continuous, hence $J(\sigma, \eta)$ is positive semi-definite for all $(\sigma, \eta) \in D(g)$ if and only if it is positive semi-definite for all $(\sigma, \eta) \in D(g)$ with $\tau = P_0\sigma \neq 0$. Using (8.2.7), we obtain for $(\kappa, \zeta) \in \mathcal{S}^3 \times \mathbb{R}$ and all such (σ, η)

$$\begin{pmatrix} \kappa \\ \zeta \end{pmatrix} \cdot \left[J(\sigma, \eta) \begin{pmatrix} \kappa \\ \zeta \end{pmatrix} \right] = \begin{pmatrix} \kappa \\ \zeta \end{pmatrix} \cdot \left[\nabla g(\sigma, \eta) \begin{pmatrix} I & 0 \\ 0 & \frac{1}{a} \end{pmatrix} \begin{pmatrix} \kappa \\ \zeta \end{pmatrix} \right] = \begin{pmatrix} \kappa \\ \zeta \end{pmatrix} \cdot \left[\nabla g(\sigma, \eta) \begin{pmatrix} \kappa \\ \frac{\zeta}{a} \end{pmatrix} \right]$$

$$= \kappa \cdot (\nabla_\sigma g_1)\kappa + \kappa \cdot (\frac{\partial}{\partial \eta} g_1) \frac{\zeta}{a} + \zeta(\nabla_\sigma g_2)\kappa + \zeta(\frac{\partial}{\partial \eta} g_2) \frac{\zeta}{a}$$

$$= \frac{F}{|\tau|} \kappa \cdot \left[I - \frac{\tau\tau^T}{|\tau|^2} \right] \kappa + \frac{F'}{-\eta} \kappa \cdot \left(\frac{\tau\tau^T}{|\tau|^2} \kappa \right) + F' \frac{|\tau|}{\eta^2} \kappa \cdot \frac{\tau}{|\tau|} \frac{\zeta}{a} + \qquad (8.2.11)$$

$$+ \gamma \left[F' \frac{|\tau|}{-\eta} + F \right] \frac{\tau^T}{|\tau|} \kappa\zeta + \frac{\zeta^2}{a} \left[F' \frac{|\tau|^2}{\eta^2} \gamma - F|\tau|\gamma' + \delta_A' \right]$$

$$= \frac{F}{|\tau|} \left(|\kappa|^2 - (\frac{\tau}{|\tau|} \cdot \kappa)^2 \right) + \begin{pmatrix} \frac{\tau}{|\tau|} \cdot \kappa \\ \zeta \end{pmatrix} \cdot \left[J^*(\sigma, \eta) \begin{pmatrix} \frac{\tau}{|\tau|} \cdot \kappa \\ \zeta \end{pmatrix} \right]$$

with

$$J^*(\sigma, \eta) = \begin{pmatrix} \frac{F'}{-\eta} & \frac{1}{a} F' \frac{|\tau|}{\eta^2} \\ (F' \frac{|\tau|}{-\eta} + F)\gamma & \frac{1}{a}(F' \frac{|\tau|^2}{\eta^2} \gamma - F|\tau|\gamma' + \delta_A') \end{pmatrix}.$$

From (8.2.11) it follows that $J(\sigma, \eta)$ is positive semi-definite if and only if $J^*(\sigma, \eta)$ is positive semi-definite. For if $J^*(\sigma, \eta)$ is positive semi-definite, then the last term of (8.2.11) is greater than or equal to zero for all (κ, ζ), since $\frac{F}{|\tau|}\left(|\kappa|^2 - (\frac{\tau}{|\tau|}\cdot\kappa)^2\right) \geq 0$. This means that $J(\sigma, \eta)$ is positive semi-definite. Assume on the other hand that $J(\sigma, \eta)$ is positive semi-definite, and let $(\xi, \zeta) \in \mathbb{R}^2$. Set $\lambda = \xi/|\tau|$ and $\kappa = \lambda\tau$. Then

$$|\kappa|^2 - \left(\frac{\tau}{|\tau|}\cdot\kappa\right)^2 = 0, \qquad \frac{\tau}{|\tau|}\cdot\kappa = \lambda|\tau| = \xi,$$

and therefore (8.2.11) implies

$$\begin{pmatrix}\xi\\\zeta\end{pmatrix}\cdot\left[J^*(\sigma, \eta)\begin{pmatrix}\xi\\\zeta\end{pmatrix}\right] = \begin{pmatrix}\kappa\\\zeta\end{pmatrix}\cdot\left[J(\sigma, \eta)\begin{pmatrix}\kappa\\\zeta\end{pmatrix}\right] \geq 0,$$

which shows that $J^*(\sigma, \eta)$ is positive semi-definite. It thus remains to study under what conditions $J^*(\sigma, \eta)$ is positive semi-definite. A 2×2-matrix $A = (a_{ij})_{i,j=1,2}$ is positive semi-definite if and only if

$$4a_{11}a_{22} \geq (a_{12} + a_{21})^2 \quad \text{and} \quad a_{11} \geq 0.$$

If we replace the elements a_{ij} by the elements of the matrix $J^*(\sigma, \eta)$ in these inequalities and note that $a_{11} = F'/(-\eta) \geq 0$ for all $(\sigma, \eta) \in D(g) = -[\mathcal{S}^3 \times (D(\gamma)\cap D(\delta_A))]$, we obtain that this matrix is positive semi-definite if and only if (8.2.1) holds.

Therefore, from what we proved, conditions (ii) and (iii) of Lemma 3.4.1 hold if and only if M is of the form (8.2.10) and (8.2.1) is satisfied.

Moreover, for M of the form (8.2.10), conditions (iv) and (8.2.2) are equivalent, since (8.1.17), (8.1.13) and (8.1.14) imply for all $(\sigma, \eta) \in D_0(f) = \mathcal{S}^3 \times (D(\gamma)\cap D(\delta_A))$

$$f(0, \sigma, \eta)\cdot M(\sigma, \eta) = g\left(-\hat{L}(\sigma, \eta)\right)\cdot M(\sigma, \eta)$$

$$= -F\frac{P_0 D\sigma}{|P_0 D\sigma|}\cdot D\sigma + (\gamma F|P_0 D\sigma| - \delta_A)\cdot a\eta = F|P_0 D\sigma|(-1 + a\eta\gamma) - a\delta_A\eta,$$

where the argument of γ and δ_A is η. Finally for M of the form (8.2.10), condition (i) is satisfied, since

$$(\kappa, \zeta)\cdot L(\kappa, \zeta) = (\kappa, \zeta)\cdot[M - B^T DB](\kappa, \zeta) = (\kappa, \zeta)\cdot(0, a\zeta) = a\zeta^2 \geq 0.$$

In summary, conditions (i) – (iv) of Lemma 3.4.1 are satisfied if and only if (8.2.1) and (8.2.2) hold. This lemma now yields that the constitutive equations (8.1.16), (8.1.17) are of monotone type if and only if (8.2.1) and (8.2.2) hold. This completes the proof of (I).

To verify (II), we use that $F(s) = F_2(s) = ds^n$ with $d > 0$, $n > 1$ satisfies

$$F'(s) = \frac{n}{s}F(s).$$

If we set $s = |\tau|/\eta$, then with this relation and with $\delta_A = 0$ the inequality (8.2.1) reduces to

$$F(s)^2\frac{4}{a}\left[n^2\frac{\gamma(\eta)}{\eta} - n\gamma'(\eta)\right] \geq F(s)^2\left[\frac{n}{a\eta} + (n+1)\gamma(\eta)\right]^2,$$

or, equivalently,

$$\frac{4n}{a}\gamma' + \frac{4n}{a\eta}\gamma + \left((n+1)\gamma - \frac{n}{a\eta}\right)^2 \leq 0,$$

which can also be written in the form (8.2.3). Consequently, (8.2.3) is equivalent to (8.2.1). The proof of Proposition 8.2.1 is complete.

8.3 Pre-Monotone Type Preserving Transformations of the Model

If the constitutive equations are not of monotone type, we can apply a transformation of interior variables. Since equations of monotone type are also of pre-monotone type, we first use Theorems 6.1.2 and 6.1.3 to study the form of transformation fields which can be used to transform the constitutive equations (8.1.2) – (8.1.4) or (8.1.16), (8.1.17) to pre-monotone type.

Proposition 8.3.1 Let $F'(s) > 0$ for $s > 0$, let $H^* : S^3 \times \mathbb{R} \to S^3 \times \mathbb{R}$ be a transformation field, and let $M^* : S^3 \times \mathbb{R} \to S^3 \times \mathbb{R}$ be a symmetric, positive definite linear mapping. Then H^* transforms the constitutive equations (8.1.16), (8.1.17) to constitutive equations of pre-monotone type, whose associated free energy ψ_{H^*} satisfies $\rho\nabla_h\psi_{H^*} \cdot (0, h) = M^*h$, if and only if an invertible linear map $A : S^3 \times \mathbb{R} \to S^3 \times \mathbb{R}$ and a constant $a > 0$ exist such that $H = AH^*$ is of the form

$$H(\tau, \zeta) = (\tau, \tilde{H}(\zeta)) \tag{8.3.1}$$

for all $(\tau, \zeta) \in S^3 \times (D(\gamma) \cap D(\delta_A))$ with a suitable mapping $\tilde{H} : D(\gamma) \cap D(\delta_A) \to \mathbb{R}$, and such that M^* is of the form $M^* = A^T M A$ with the symmetric, positive definite linear mapping

$$M(\tau, \zeta) = (D\tau, a\zeta), \quad (\tau, \zeta) \in S^3 \times \mathbb{R}. \tag{8.3.2}$$

The vector field H transforms (8.1.16), (8.1.17) to constitutive equations of pre-monotone type, whose associated free energy ψ_H satisfies $\rho\nabla_h\psi_H(0, h) = Mh$.

Proof: Since by Lemma 8.2.2

$$\bigcap_{\varepsilon \in D(f, z)} \ker[B\nabla g(\overline{L}\varepsilon - \hat{L}z)] = \{0\}$$

for all $z = (\tau, \zeta) \in S^3 \times \mathbb{R} \cong \mathbb{R}^7$ with $D(f, z) \neq \emptyset$, it follows that condition (6.1.23) is satisfied. Theorem 6.1.3 thus implies that if H^* transforms (8.1.16), (8.1.17) to pre-monotone type, then an invertible linear map $A : S^3 \times \mathbb{R} \to S^3 \times \mathbb{R}$ exists such that $H = AH^*$ is of the form

$$H(z) = Pz + \tilde{H}(Qz) \tag{8.3.3}$$

on $D_0(f) + R(\hat{L}^{-1}\overline{L}) = -\hat{L}^{-1}D(g) = S^3 \times (D(\gamma) \cap D(\delta_A))$, with projectors $P : S^3 \times \mathbb{R} \to R(\hat{L}^{-1}\overline{L})$ along $\ker(B)$ and $Q : S^3 \times \mathbb{R} \to \ker(B)$ along $R(\hat{L}^{-1}\overline{L})$, and with $\tilde{H} : \ker(B) \to \ker(B)$. If we use that (8.1.11) – (8.1.13) imply

$$R(\hat{L}^{-1}\overline{L}) = S^3 \times \{0\}, \quad \ker(B) = \{0\} \times \mathbb{R},$$

and if we identify $\tilde{H}$: $\ker(B) \to \ker(B)$ with a real function $\tilde{H}$: $\mathbb{R} \to \mathbb{R}$, we immediately see that (8.3.1) and (8.3.3) are equivalent, whence A must exist such that $H = AH^*$ satisfies (8.3.1). Theorem 6.1.3 also implies that $H = AH^*$ transforms (8.1.16), (8.1.17) to equations of pre-monotone type, for which the gradient $Mh = \rho\nabla_h\psi_H(0,h)$ of the associated free energy ψ_H satisfies $M^{-1}\overline{M} = \hat{L}^{-1}\overline{L}$. Using $\overline{M} = \overline{L} = B^T D$, this is equivalent to $(\hat{L}M^{-1} - I)\overline{L} = 0$. This equation implies (8.3.2), which is seen just as in the proof of Lemma 8.2.3. Since $H = AH^*$, the transformation performed by the vector field H is obtained by composition of the transformations performed by A and by H^*. Therefore we can apply Lemma 5.2.2 and obtain from (5.2.7) that $M = (A^{-1})^T M^* A^{-1}$. Whence $M^* = A^T M A$, and the first part of the proof is complete.

To prove the converse we use Theorem 6.1.2. Assume that a linear mapping M and a continuously differentiable vector field H^* : $S^3 \times \mathbb{R} \to S^3 \times \mathbb{R}$ are given, such that M satisfies (8.3.2), and such that an invertible linear map A exists, for which $H = AH^*$ satisfies (8.3.1) on the set $S^3 \times (D(\gamma) \cap D(\delta_A))$. Then M is symmetric, positive definite, and satisfies conditions (i) and (ii) of Theorem 6.1.2. Moreover, H^* can be redefined outside the set $S^3 \times (D(\gamma) \cap D(\delta_A))$, such that H has the form (8.3.1) and therefore the form (6.1.14) on all of $S^3 \times \mathbb{R}$. Theorem 6.1.2 thus implies that H transforms the constitutive equations (8.1.16), (8.1.17) to constitutive equations of pre-monotone type, whose associated free energy ψ_H satisfies $\rho\nabla_h\psi_H(0,h) = Mh$. Since H transforms (8.1.16), (8.1.17) into constitutive equations of pre-monotone type, we conclude from Lemma 5.2.2 that $H^* = A^{-1}H$ also transforms (8.1.16), (8.1.17) into constitutive equations of pre-monotone type, whose associated free energy ψ_{H^*} is given by $\psi_{H^*}(\varepsilon, h) = \psi_H(\varepsilon, Ah)$ and therefore satisfies $\rho\nabla_h\psi_{H^*}(0, h) = M^*h = A^T M Ah$. The proof is complete.

8.4 Transformation to Gradient Type

In the following we study the transformation of the constitutive equations (8.1.16), (8.1.17) into equations from the classes $\mathcal{M}$, $\mathcal{G}$ and $\mathcal{MG}$. Since these are all subclasses of the class $\mathcal{M}^*$ of pre-monotone type, it follows from Proposition 8.3.1 that to each vector field H^*, which transforms (8.1.16), (8.1.17) into equations from one of the classes $\mathcal{M}$, $\mathcal{G}$, $\mathcal{MG}$, there exists an invertible linear mapping A such that $H = AH^*$ is of the form (8.3.1). Since by Lemma 5.2.2 the classes $\mathcal{M}$, $\mathcal{G}$, $\mathcal{MG}$ are invariant under transformation by A, it follows that H also transforms (8.1.16), (8.1.17) to $\mathcal{M}$, $\mathcal{G}$, or $\mathcal{MG}$. Therefore, to study whether (8.1.16), (8.1.17) belong to one of the classes $\mathcal{TM}$, $\mathcal{TG}$, or $\mathcal{T}(\mathcal{MG})$, it suffices to study whether (8.1.16), (8.1.17) can be transformed to the classes $\mathcal{M}$, $\mathcal{G}$, or $\mathcal{MG}$ by a transformation field satisfying (8.3.1) on $S^3 \times (D(\gamma) \cap D(\delta_A))$.

We first use Theorem 6.2.4 to study when (8.1.16), (8.1.17) belong to $\mathcal{TG}$.

Proposition 8.4.1 *Assume that $F'(s) > 0$ for $s > 0$ and that $D(\gamma) \cap D(\delta_A)$ is an interval. Then the following assertions hold:*
The constitutive equations (8.1.16), (8.1.17) belong to the class $\mathcal{TG}$ if and only if there exists a constant $\theta > 0$ such that

$$F(s) = \theta F'(s)s \tag{8.4.1}$$

for all $s \geq 0$.

Let $a > 0$ *and let the linear mapping* $M : S^3 \times \mathbb{R} \to S^3 \times \mathbb{R}$ *be defined by* $M(\tau, \zeta) = (D\tau, a\zeta)$. *Then the constitutive equations are transformed to equations*

$$T = D(\varepsilon - Bh) \tag{8.4.2}$$

$$h_t = \nabla\chi(\overline{M}\varepsilon - Mh) \tag{8.4.3}$$

of the class $\mathcal{G}$ *by a vector field of the form (8.3.1) if and only if*

$$\tilde{H}'(\zeta)^2 = \frac{1}{a(1 + \theta)\gamma(\zeta)\zeta}, \tag{8.4.4}$$

holds for $\zeta \in D(\gamma) \cap D(\delta_A)$.

Example 4. If $F = F_1$, then the constitutive equations (8.1.16) (8.1.17) do not belong to $\mathcal{TG}$, since $F'(s) = \frac{\sigma n}{s^{n+1}}F(s)$ holds and (8.4.1) is therefore not satisfied.

Example 5. If $F = F_2$, then the constitutive equations (8.1.16), (8.1.17) belong to the class $\mathcal{TG}$ and are transformed into equations of the class $\mathcal{G}$ by a vector field of the form (8.3.1) if and only if

$$\tilde{H}'(\zeta)^2 = \frac{n}{a(n + 1)\zeta\gamma(\zeta)} \tag{8.4.5}$$

holds for $\zeta \in D(\gamma) \cap D(\delta_A)$. This follows immediately from Proposition 8.4.1, since $F(s) = F_2(s) = ds^n$ satisfies (8.4.1) with $\theta = 1/n$.

Proof of Proposition 8.4.1: To apply Theorem 6.2.4 we need to determine the coordinates $(\hat{z}, \tilde{z}) = (\hat{z}_1, \ldots, \hat{z}_d, \tilde{z}_1, \ldots, \tilde{z}_{N-d})$ and the sets $\hat{D}_0(f, \tilde{z})$, $\tilde{D}_0(f, \hat{z})$ used in the formulation of this theorem. Now, since we identified S^3 with $\mathbb{R}^6$, since $N = 7$, and since (8.1.11) – (8.1.13) imply

$$R(\hat{L}^{-1}\overline{L}) = S^3 \times \{0\}, \quad \ker(B) = \{0\} \times \mathbb{R},$$

hence $d = 6$, we obtain $\tilde{z} \in \mathbb{R}$, and for $\hat{z}_1, \ldots, \hat{z}_6$ we can choose the coordinates in the six-dimensional manifold S^3 defined by the isomorphism (3.3.1). Moreover,

$$D_0(f) = \{z \in S^3 \times \mathbb{R} \mid (0, z) \in D(f)\} = -\hat{L}^{-1}D(g) = S^3 \times (D(\gamma) \cap D(\delta_A))$$

yields

$$\hat{D}_0(f, \tilde{z}) = \{\hat{z} \in S^3 \mid (\hat{z}, \tilde{z}) \in D_0(f)\} = S^3 \quad \text{or} \quad \hat{D}_0(f, \tilde{z}) = \emptyset$$

$$\tilde{D}_0(f, \hat{z}) = \{\tilde{z} \in \mathbb{R} \mid (\hat{z}, \tilde{z}) \in D_0(f)\} = D(\gamma) \cap D(\delta_A).$$

Consequently, since by assumption $D(\gamma) \cap D(\delta_A)$ is an interval and therefore connected, conditions (i) and (ii) of Theorem 6.2.4 are satisfied if we set $\hat{z}_0 = 0$ and choose for Γ the identity on S^3. In Lemma 8.2.2 we showed that condition (6.2.24) is satisfied. Theorem 6.2.4 (III) together with Proposition 8.3.1 thus imply that the constitutive equations (8.1.16), (8.1.17) belong to the class $\mathcal{TG}$ if and only if there exist a linear mapping M of the form $M(\hat{z}, \tilde{z}) = (D\hat{z}, a\tilde{z})$ for a constant $a > 0$ and a function $\tilde{H} : \mathbb{R} \to \mathbb{R}$ which satisfy conditions (iii) and (iv) of Theorem 6.2.4. The function $\tilde{H}$ will then define the transformation field H via (8.3.1).

To study the conditions (iii), (iv) of Theorem 6.2.4, we note that for such M

$$\hat{M} = P^{\perp}M|_{R(\hat{L}^{-1}\hat{L})} = D, \quad \check{M} = Q^{\perp}M|_{\ker(B)} = a.$$ (8.4.6)

Since

$$D\frac{\partial}{\partial \hat{z}_i}\hat{f}(0, \hat{z}, \tilde{z}) = \frac{\partial}{\partial \hat{z}_i}D\hat{f}(0, \hat{z}, \tilde{z}),$$

condition (iii) means that the Jacobi-matrix $\nabla_{\hat{z}}(D\hat{f})$ must be symmetric. From (8.1.17) and (8.1.14) we obtain with $\hat{w} = P_0 D\hat{z}$,

$$\hat{f}(0, \hat{z}, \tilde{z}) = g_1(-\hat{L}(\hat{z}, \tilde{z})) = g_1(-D\hat{z}, -\tilde{z}) = -F(|\hat{w}|/\tilde{z})\hat{w}/|\hat{w}|.$$

By our convention of writing gradients as column vectors, for a primitive G of $-F$ we thus obtain that

$$\nabla_{\hat{z}}G(|P_0 D\hat{z}|/\tilde{z})\tilde{z} = \left[-F(|\hat{w}|/\tilde{z})(\hat{w}/|\hat{w}|)^T P_0 D\right]^T = D^T P_0^T g_1 = Dg_1 = D\hat{f},$$

where we used that the orthogonal projection P_0 satisfies $P_0^T = P_0$, and that $P_0 g_1 = g_1$, which follows immediately from (8.1.14). This implies that $\nabla_{\hat{z}}(D\hat{f})$ is the Hessian matrix of $\hat{z} \mapsto G(|P_0 D\hat{z}|/\tilde{z})\tilde{z}$, and as a consequence, is symmetric. Condition (iii) is therefore satisfied and (8.1.16), (8.1.17) belong to $T\mathcal{G}$ if and only if condition (iv) holds.

This condition consists of the two equations (6.2.22), (6.2.23). Equation (6.2.23) is automatically satisfied, since $\tilde{z}$ is a real variable and thus has only one component. To study (6.2.22), we note that since Γ is the identity, the vector field $\omega : S^3 \setminus \{0\} \to S^3$ defined in (6.2.19) is given by

$$\omega(\hat{z}) = \hat{z}/|\hat{z}|.$$

Thus, if we set $\hat{w} = P_0 D\hat{z}$, $\eta = |\hat{w}|/\tilde{z}$, then together with $g_1 = P_0 g_1$ we obtain for the right hand side of (6.2.22) that

$$\frac{\partial}{\partial \tilde{z}}(\omega(\hat{z}) \cdot \hat{M}\hat{f}(0, \hat{z}, \tilde{z}))$$

$$= \frac{\hat{z}}{|\hat{z}|} \cdot D\frac{\partial}{\partial \tilde{z}}g_1(-\hat{L}(\hat{z}, \tilde{z})) = \frac{\hat{z}}{|\hat{z}|} \cdot DP_0 \frac{\partial}{\partial \tilde{z}}g_1(-\hat{L}(\hat{z}, \tilde{z}))$$ (8.4.7)

$$= \frac{\hat{w}}{|\hat{z}|} \cdot F'(|\hat{w}|/\tilde{z})\frac{\hat{w}}{|\hat{w}|}\frac{|\hat{w}|}{\tilde{z}^2} = F'(\eta)\eta^2/|\hat{z}|.$$

To compute the left hand side of (6.2.22), we use that (8.1.17) and (8.1.14) yield

$$(\omega(\hat{z}) \cdot \nabla_{\hat{z}})\hat{f}(0, \hat{z}, \tilde{z}) = (\omega(\hat{z}) \cdot \nabla_{\hat{z}})g_2(-\hat{L}(\hat{z}, \tilde{z}))$$

$$= \frac{\hat{z}}{|\hat{z}|} \cdot \nabla_{\hat{z}}[\gamma(\tilde{z})F(|\hat{w}|/\tilde{z})|\hat{w}| - \delta_A(\tilde{z})]$$

$$= \gamma(\tilde{z})\left[\frac{F'(\eta)}{\tilde{z}}|\hat{w}| + F(\eta)\right]\frac{\hat{z}}{|\hat{z}|} \cdot \nabla_{\hat{z}}|P_0 D\hat{z}|$$

$$= \tilde{z}\gamma(\tilde{z})(F'(\eta)\eta + F(\eta))\eta/|\hat{z}|.$$

With this equation and with (8.4.6), (8.4.7), condition (6.2.22) reduces to

$$a\left(\hat{z}\gamma(\tilde{z})(F'(\eta)\eta + F(\eta))\eta/|\hat{z}|\right)\tilde{H}'(\tilde{z})^2 = F'(\eta)\eta^2/|\hat{z}|\,.$$

After simplification we obtain the equation

$$a\left(1 + \frac{F(\eta)}{F'(\eta)\eta}\right)\tilde{H}'(\tilde{z})^2 = \frac{1}{\gamma(\tilde{z})\tilde{z}}\,,$$

which must hold for all $(\hat{z}, \tilde{z}) \in D_0(f) = S^3 \times (D(\gamma) \cap D(\delta_A))$, whence for all $\eta = |P_0 D\hat{z}|/\tilde{z} \geq 0$ and all $\tilde{z} \in D(\gamma) \cap D(\delta_A)$. It is immediately seen that this equation holds, hence condition (iv) of Theorem 6.2.4 is satisfied and consequently (8.1.16), (8.1.17) belong to $\mathcal{TG}$ if and only if (8.4.1) and (8.4.4) are satisfied. The proof is complete.

8.5 Transformation to Monotone Type and to Monotone-Gradient Type

We next determine when (8.1.16), (8.1.17) belong to the classes $\mathcal{TM}$ and $\mathcal{T(MG)}$.

Proposition 8.5.1 *Assume that* $F(0) = F'(0) = 0$, $F'(s) > 0$ *for* $s > 0$ *and that* $D(\gamma) \cap D(\delta_A)$ *is an interval. Then the following assertions hold:*
(I) The constitutive equations (8.1.16), (8.1.17) belong to the class $\mathcal{TM}$ *if and only if a function* $\tilde{H} : \mathbb{R} \to \mathbb{R}$ *and a constant* $a > 0$ *exist such that the inequalities*

$$\frac{4}{a}\frac{F'}{\eta}\left[F'\frac{|\tau|^2}{\eta^2}\gamma - F|\tau|\gamma' + \delta_A' - (\gamma F|\tau| - \delta_A)\frac{\tilde{H}''}{\tilde{H}'}\right]$$

$$\geq \left[\frac{1}{a\tilde{H}'}F'\frac{|\tau|}{\eta^2} + (F'\frac{|\tau|}{\eta} + F)\gamma\tilde{H}'\right]^2 \tag{8.5.1}$$

$$a\gamma(\eta)\tilde{H}'(\eta)\tilde{H}(\eta) \leq 1\,, \quad \delta_A(\eta)\tilde{H}'(\eta)\tilde{H}(\eta) \geq 0 \tag{8.5.2}$$

hold for all $(\tau, \eta) = (P_0\sigma, \eta)$ *with* $(\sigma, \eta) \in -D(g) = S^3 \times (D(\gamma) \cap D(\delta_A))$. *In (8.5.1) the argument of* F *and* F' *is* $|\tau|/\eta$, *and the argument of* γ, δ_A, $\tilde{H}'$, $\tilde{H}''$ *is* η.
    If $\tilde{H}$ *and* a *satisfy (8.5.1), (8.5.2), then the vector field* $H(\sigma, \eta) = (\sigma, \tilde{H}(\eta))$ *transforms (8.1.16), (8.1.17) to constitutive equations of the class* $\mathcal{M}$, *whose associated free energy* ψ_H *satisfies* $\rho\nabla_h\psi_H(0, (\sigma, \eta)) = M(\sigma, \eta) = (D\sigma, a\eta)$ *for all* $(\sigma, \eta) \in S^3 \times \mathbb{R}$.
(II) If $F = F_2$ *and* $\delta_A = 0$, *then (8.5.1) is equivalent to*

$$\eta\left[(\eta\gamma)' + \eta\gamma\frac{\tilde{H}''}{\tilde{H}'}\right] \leq -\frac{a}{4n}\left((n+1)\eta\gamma\tilde{H}' - \frac{n}{a\tilde{H}'}\right)^2 \tag{8.5.3}$$

for all $\eta \in D(\gamma)$, *where the argument of* γ, $\tilde{H}'$, $\tilde{H}''$ *is* η.
(III) Let $\delta_A = 0$. *Then the constitutive equations (8.1.16), (8.1.17) belong to the class* $\mathcal{T(MG)}$ *if and only if the two relations*

$$F(s) = \theta F'(s)s\,, \quad s \geq 0\,, \tag{8.5.4}$$

with a suitable constant $\theta > 0$, and

$$\frac{d}{d\eta}[\eta\gamma(\eta)] \le 0, \quad \eta \in D(\gamma), \tag{8.5.5}$$

hold.

Proof. Since according to Proposition 8.3.1 and Lemma 5.2.2 the constitutive equations (8.1.16), (8.1.17) belong to $\mathcal{T M}$ if and only if they can be transformed to monotone type by a vector field of the form (8.3.1), it suffices to determine when the constitutive equations can be transformed to equations of the class $\mathcal{M}$ by such a vector field.

Thus, let H be a transformation field of the form (8.3.1) and let M be the linear mapping defined by $Mh = \rho\nabla_h\psi_H(0,h)$, where ψ_H is the free energy associated to the transformed constitutive equations of pre-monotone type. Then, according to Proposition 8.3.1, M is of the form $M(\sigma,\eta) = (D\sigma, a\eta)$, and according to Theorem 5.3.1 and its proof, if H satisfies condition (vii) of this theorem, then the transformed constitutive equations belong to the class $\mathcal{M}$ if and only if H and M satisfy the conditions (v) and (vi) of this theorem. Now, for every transformation field of the form (8.3.1), the set

$$H(D_0(f)) = H(-\hat{L}^{-1}D(g)) = H(\hat{L}^{-1}[\mathcal{S}^3 \times (D(\gamma) \cap D(\delta_A))]) = \mathcal{S}^3 \times \tilde{H}(D(\gamma) \cap D(\delta_A))$$

is convex. This holds since we assumed that $D(\gamma) \cap D(\delta_A)$ is an interval and since, as we always assume, $\tilde{H}$ is continuously differentiable and therefore continuous, which implies that $\tilde{H}(D(\gamma) \cap D(\delta_A))$ is also an interval. Consequently, condition (vii) is satisfied, and to prove (I) it remains to show that the inequalities (8.5.1), (8.5.2) hold if and only if conditions (v) and (vi) of Theorem 5.3.1 are satisfied.

We first show that the matrix

$$\mathcal{A}_H(z) = H'(z)\nabla_z f(0,z)H'(z)^{-1}M^{-1} + H''(z) : f(0,z)H'(z)^{-1}M^{-1}$$

is negative semi–definite if and only if (8.5.1) holds. For H of the form (8.3.1) the tensor $H''(z)$ defined in Lemma 5.2.1 satisfies

$$H''(z) : f(0,z) = H''(z) : g(-\hat{L}z) = H''(\tau,\zeta) : g(-D\sigma, -\eta)$$
$$= \begin{pmatrix} 0 & 0 \\ 0 & \tilde{H}''(\eta)g_2(-D\sigma, -\eta) \end{pmatrix},$$

where we set $z = (\sigma,\eta) \in \mathcal{S}^3 \times (D(\gamma) \cap D(\delta_A))$. Using (8.3.1) and $M(\sigma,\eta) = (D\sigma, a\eta)$, after a short computation we obtain that

$$\mathcal{A}_H(z) = H'(z)\nabla g(-\hat{L}z)(-\hat{L})H'(z)^{-1}M^{-1} + H''(z) : g(-\hat{L}z)H'(z)^{-1}M^{-1}$$
$$= -\begin{pmatrix} \nabla_\sigma g_1 & \frac{1}{a\tilde{H}'}\frac{\partial}{\partial\eta}g_1 \\ \tilde{H}'\nabla_\sigma g_2 & \frac{1}{a}(\frac{\partial}{\partial\eta}g_2 - \frac{\tilde{H}''}{\tilde{H}'}g_2) \end{pmatrix}, \tag{8.5.6}$$

where the argument of g_1 and g_2 is $(-D\sigma, -\eta)$, and where the argument of $\tilde{H}', \tilde{H}''$ is η. The matrix on the right hand side of (8.5.6) has exactly the same structure as the matrix $J(\sigma,\eta)$ in the proof of Proposition 8.2.1. By the same computation and the same arguments as in that proof we thus obtain that $A_H(\sigma, \eta)$ is negative semi-definite for all $(\sigma, \eta) \in D_0(f)$, and therefore that condition (v) holds if and only if inequality (8.5.1) is satisfied.

To show that condition (vi) holds if and only if (8.5.2) is satisfied, we use that (8.1.17), (8.1.14), (8.3.1) and $M(\sigma, \eta) = (D\sigma, a\eta)$ together yield for $z = (\sigma, \eta)$ and $w = P_0 D\sigma$

$$(H'(z)f(0, z)) \cdot (MH(z)) = (H'(z)g(-\hat{L}z)) \cdot (MH(z))$$

$$= -F\left(\frac{|w|}{\eta}\right)\frac{w}{|w|} \cdot D\sigma + \tilde{H}'(\eta)\left[\gamma(\eta)F\left(\frac{|w|}{\eta}\right)|w| - \delta_A(\eta)\right]a\tilde{H}(\eta)$$

$$= F|w|(a\gamma\tilde{H}'\tilde{H} - 1) - a\delta_A\tilde{H}'\tilde{H}, \qquad (8.5.7)$$

where we used that $w \cdot D\sigma = P_0 D\sigma \cdot D\sigma = |P_0 D\sigma|^2 = |w|^2$, since P_0 is an orthogonal projection. However, since we assume that $F'(s) > 0$ for $s > 0$, the function $F\left(\frac{|w|}{\eta}\right)|w|$ takes every non-negative real value as σ runs over $\mathcal{S}^3$. This implies that the expression on the right hand side of (8.5.7) is non-positive for all $(\sigma, \eta) \in D_0(f) = \mathcal{S}^3 \times (D(\gamma) \cap D(\delta_A))$, and thus that condition (vi) of Theorem 5.3.1 is satisfied if and only if the inequalities (8.5.2) hold. This completes the proof of (I).

To prove (II), we use that $F(s) = F_2(s) = ds^n$ satisfies

$$F'(s) = \frac{n}{s}F(s).$$

If we insert this equation into (8.5.1), use that $\delta_A = 0$, and set $s = |\tau|/\eta$, we obtain

$$\frac{4}{a\eta}\left[n^2\gamma - n\eta\gamma' - n\eta\gamma\frac{\tilde{H}''}{\tilde{H}'}\right] \geq \left[\frac{n}{a\eta\tilde{H}'} + (n+1)\gamma\tilde{H}'\right]^2.$$

With $n^2\gamma - n\eta\gamma' = n(n+1)\gamma - n(\eta\gamma)'$ this can also be written in the form

$$-\frac{4n}{a\eta}\left[(\eta\gamma)' + \eta\gamma\frac{\tilde{H}''}{\tilde{H}'}\right] \geq \left[\frac{n}{a\eta\tilde{H}'} - (n+1)\gamma\tilde{H}'\right]^2,$$

and this is equivalent to (8.5.3).

To prove (III), we note that in view of Proposition 8.4.1, the constitutive equations (8.1.16), (8.1.17) belong to $\mathcal{TG}$ if and only if (8.5.4) is satisfied. Moreover, every vector field $H(\tau, \varsigma) = (\tau, \tilde{H}(\varsigma))$ which transforms these constitutive equations to gradient type is a solution of (8.4.4). Above we proved that $H(D_0(f))$ is convex. Therefore the domain $D(\chi) = -MH(D_0(f))$ of the function χ from the transformed equation (8.4.3) is also convex, and this implies that the vector field $\nabla\chi$ is monotone if and only if χ is convex. By Definition 3.1.1 this means that the transformed equations (8.4.2), (8.4.3) belong to the class $\mathcal{MG}$ if and only if they are of monotone type. Thus, to prove statement (III), it suffices to show that (8.5.5) holds if and only if a solution of (8.4.4) can be chosen for which the transformed equations (8.4.2), (8.4.3) are of monotone type.

To show this, we use that (8.5.4), $F'(s) > 0$ for $s > 0$ and $F'(0) = 0$ together imply $F(s) = F_2(s) = ds^n$ with $n = 1/\theta > 1$. From the first two parts of this proof we thus obtain that the transformed equations are of monotone type if and only if (8.5.2) and (8.5.3) hold. We thus complete the proof by showing that if (8.4.4) holds, then firstly, (8.5.5) and (8.5.3) are equivalent, and secondly, if (8.5.5) holds then initial conditions for the solution $\tilde{H}$ of the differential equation (8.4.4) can be chosen such that (8.5.2) holds.

To prove the equivalence of (8.5.3) and (8.5.5), note that with $n = 1/\theta$ the equation (8.4.4) takes the form (8.4.5). Equation (8.4.5) implies that the right hand side of (8.5.3) vanishes, and that

$$\frac{\tilde{H}''}{\tilde{H}'} = \frac{\tilde{H}'\tilde{H}''}{(\tilde{H}')^2} = \frac{1}{2}\frac{\frac{d}{d\eta}(\tilde{H}')^2}{(\tilde{H}')^2} = \frac{1}{2}\eta\gamma\frac{d}{d\eta}(\eta\gamma)^{-1} = -\frac{1}{2}\frac{(\eta\gamma)'}{\eta\gamma},$$

which reduces (8.5.3) to

$$\frac{1}{2}\eta(\eta\gamma)' \le 0.$$

This inequality must hold for all $\eta \in D(\gamma)$. Since we assumed in (8.1.5) that $D(\gamma) \subseteq \mathbb{R}^+$, this inequality and consequently also (8.5.3) is equivalent to (8.5.5).

To deduce that if (8.5.5) holds, then (8.5.2) is satisfied after a suitable choice of the initial condition for the solution $\tilde{H}$ of (8.4.4), let

$$S(\eta) = \sqrt{\frac{(1+\theta)}{a}}\sqrt{\frac{\eta}{\gamma(\eta)}}, \qquad \eta \in D(\gamma) \subseteq \mathbb{R}^+.$$

Then (8.5.5) and (8.4.4) yield

$$S'(\eta) = \frac{1}{2}\sqrt{\frac{(1+\theta)}{a}}\sqrt{\frac{\gamma}{\eta}}\left[\frac{1}{\gamma} - \frac{\eta\gamma'}{\gamma^2}\right] = \frac{1}{2}\sqrt{\frac{(1+\theta)}{a}}\sqrt{\frac{\gamma}{\eta}}\left[\frac{2}{\gamma} - \frac{(\eta\gamma)'}{\gamma^2}\right]$$

$$\ge \sqrt{\frac{(1+\theta)}{a}}\sqrt{\frac{\gamma}{\eta}}\frac{1}{\gamma} = \frac{1+\theta}{\sqrt{a(1+\theta)\eta\gamma}} > \frac{1}{\sqrt{a(1+\theta)\eta\gamma}} = \tilde{H}'(\eta). \quad (8.5.8)$$

Therefore, if we choose the initial condition for the differential equation (8.4.4) at $\zeta_2 = \inf D(\gamma)$ such that $\tilde{H}(\zeta_2) \le S(\zeta_2)$, then (8.5.8) implies $\tilde{H}(\eta) \le S(\eta)$ for all $\eta \in D(\gamma)$, hence, from (8.4.4),

$$a\gamma\tilde{H}'\tilde{H} \le \frac{a\gamma}{[a(1+\theta)\gamma\eta]^{1/2}}\tilde{H} = \frac{\tilde{H}}{S} \le 1,$$

which is (8.5.2). The proof is complete.

Example 6 Let $F = F_2$, $\delta_A = 0$ and $\gamma(\eta) = \gamma_2(\eta) = m\eta^{-r}$. Then the following assertions hold:
(I) The constitutive equations (8.1.16), (8.1.17) belong to the class $\mathcal{T}(\mathcal{MG})$ and thus in particular to $\mathcal{TM}$, if and only if $r \ge 1$.
(II) For $0 < r < 1$ the equations (8.1.16), (8.1.17) are not in $\mathcal{TM}$, since they cannot be transformed such that the transformed function g_H is monotone in the whole domain

$$-MH\big(D_0(f)\big) = -\big[\mathcal{S}^3 \times a\tilde{H}\big(D(\gamma)\big)\big] = -\big[\mathcal{S}^3 \times \big[a\tilde{H}(\zeta_2), \infty\big)\big].$$

Monotonicity can only be achieved in subsets $-\big(\mathcal{S}^3 \times [\zeta_3, \zeta_4]\big)$ with $0 < \zeta_3 < \zeta_4 < \infty$ and $|\zeta_3 - \zeta_4|$ sufficiently small.

Assertion (I) should be compared to Example 2, where it is shown that (8.1.16), (8.1.17) belong to the class $\mathcal{M}$ only for $r = 1$. By transformation to gradient type the range of applicability of monotonicity methods to the equations (8.1.15) – (8.1.17) is thus enlarged to $r \geq 1$.

Proof of assertion (I). This assertion follows immediately from Proposition 8.5.1 (III), since $F = F_2$ satisfies (8.5.4) and since the inequality (8.5.5) holds for all $\eta \in D(\gamma) = [\zeta_2, \infty)$ if and only if $r \geq 1$. This is seen from

$$\frac{d}{d\eta}\left[\eta\gamma(\eta)\right] = \frac{d}{d\eta}\, m\eta^{1-r} = (1 - r)m\eta^{-r}.$$

Proof of assertion (II). We show that a and $\tilde{H}$ cannot be found such that (8.5.3) holds for all $\eta \in D(\gamma) = [\zeta_2, \infty)$. Statements (I) and (II) of Proposition 8.5.1 then together imply that the constitutive equations are not in $\mathcal{TM}$.

To this end we note that (8.5.3) can be written in the equivalent form

$$\eta\sqrt{\eta\gamma}\left[2\left(\sqrt{\eta\gamma}\right)' + \sqrt{\eta\gamma}\,\frac{\tilde{H}''}{\tilde{H}'}\right] \leq -\frac{1}{4n}\,\eta\gamma\left((n + 1)\sqrt{a\eta\gamma}\,\tilde{H}' - \frac{n}{\sqrt{a\eta\gamma}\,\tilde{H}'}\right)^2. \quad (8.5.9)$$

The left hand side is equal to

$$\eta\sqrt{\eta\gamma}\left[\left(\sqrt{\eta\gamma}\right)' + \frac{\left(\sqrt{\eta\gamma}\,\tilde{H}'\right)'}{\tilde{H}'}\right] = \eta\,\frac{\sqrt{\eta\gamma}}{\sqrt{a}\,\tilde{H}'}\left[\left(\sqrt{a\eta\gamma}\right)'\tilde{H}' + \left(\sqrt{a\eta\gamma}\,\tilde{H}'\right)'\right]$$

$$= \eta\,\frac{\sqrt{\eta\gamma}}{\sqrt{a}\,\tilde{H}'}\left[\frac{1}{2}\,\frac{(\eta\gamma)'}{\eta\gamma}\,\sqrt{a\eta\gamma}\,\tilde{H}' + \left(\sqrt{a\eta\gamma}\,\tilde{H}'\right)'\right].$$

Insertion into (8.5.9) yields

$$\left(\sqrt{a\eta\gamma}\,\tilde{H}'\right)' \leq -\frac{1}{4n\eta}\,\sqrt{a\eta\gamma}\,\tilde{H}'\left((n + 1)\sqrt{a\eta\gamma}\,\tilde{H}' - \frac{n}{\sqrt{a\eta\gamma}\,\tilde{H}'}\right)^2 - \frac{1}{2}\,\frac{(\eta\gamma)'}{\eta\gamma}\,\sqrt{a\eta\gamma}\,\tilde{H}'.$$

With the abbreviation

$$Y(\eta) = \sqrt{a\eta\gamma(\eta)}\,\tilde{H}'(\eta), \quad (8.5.10)$$

this can be written in the form

$$Y' \leq -\frac{1}{4n\eta}\left[Y\left((n + 1)Y - \frac{n}{Y}\right)^2 + 2n\,\frac{(\eta\gamma)'}{\gamma}\,Y\right], \quad \eta \in D(\gamma). \quad (8.5.11)$$

Equation (8.5.10) implies $Y(\eta) > 0$. Therefore (8.5.11) has a solution $Y : D(\gamma) \to \mathbb{R}^+$ if and only if (8.5.3) holds.

We insert $\gamma(\eta) = \gamma_2(\eta) = m\eta^{-r}$ with $r < 1$ into the inequality (8.5.11). With $(\eta\gamma_2)'/\gamma_2 = 1 - r > 0$, this inequality reduces to

$$Y' \leq -\frac{1}{4n\eta}\,\frac{1}{Y}\,p_{1-r}(Y^2), \quad (8.5.12)$$

where p_c denotes the polynomial

$$p_c(\lambda) = (n + 1)^2\lambda^2 - 2n(n + 1 - c)\lambda + n^2. \quad (8.5.13)$$

Using the following lemma we show that (8.5.12) and therefore (8.5.11) does not have a solution $Y : D(\gamma_2) = [\zeta_2, \infty) \to \mathbb{R}^+$.

Lemma 8.5.2 *Let* $0 < c < 2n + 2$, $\kappa = n/(n+1)^2$, $\omega = \sqrt{c(2n+2-c)}$ *and* $0 < \eta_1 < \eta_2$. *Suppose that* $y : [\eta_1, \eta_2] \to \mathbb{R}^+$ *satisfies*

$$y'(\eta) = -\frac{1}{4n\eta} \frac{1}{y(\eta)} p_c(y(\eta)^2). \tag{8.5.14}$$

Then

$$y(\eta_1)^2 = \kappa\omega \tan\left[\arctan\left(\frac{1}{\kappa\omega}\left(y(\eta_2)^2 - \kappa(n+1-c)\right)\right) + \frac{\omega}{2}\ln\frac{\eta_2}{\eta_1}\right] + \kappa(n+1-c), \tag{8.5.15}$$

whence

$$y(\eta_1) \to \infty \quad \text{for} \quad \eta_1 \searrow c_1\eta_2, \tag{8.5.16}$$

where

$$c_1 = \exp\left\{-\frac{2}{\omega}\left[\frac{\pi}{2} - \arctan\left(\frac{1}{\kappa\omega}\left(y(\eta_2)^2 - \kappa(n+1-c)\right)\right)\right]\right\}.$$

c_1 *satisfies* $\exp(-2\pi/\omega) < c_1 < 1$.

Proof of Lemma 8.5.2. Since

$$[2n(n+1-c)]^2 - 4(n+1)^2n^2 = 4n^2c(c-2n-2) < 0$$

for $0 < c < 2n+2$, it follows that the polynomial $p_c(\lambda)$ defined in (8.5.13) does not have real zeros, hence (8.5.14) can be written in the form

$$\frac{2yy'}{p_c(y^2)} = -\frac{1}{2n\eta}. \tag{8.5.17}$$

With the primitive

$$G(\lambda) = \frac{1}{n\omega}\arctan\left(\frac{1}{\kappa\omega}\left(\lambda - \kappa(n+1-c)\right)\right)$$

of $1/p_c(\lambda)$ equation (8.5.17) obtains the form

$$\frac{d}{d\lambda}G\left(y(\eta)^2\right) = -\frac{1}{2n\eta}.$$

This implies

$$G\left(y(\eta_1)^2\right) = G\left(y(\eta_2)^2\right) + \frac{1}{2n}\ln\frac{\eta_2}{\eta_1},$$

which after a short computation yields (8.5.15).

End of the proof of assertion (II). Let Y be a solution of (8.5.12) in $[\eta_1, \eta_2]$, let y be a solution of (8.5.14) in $[\eta_1, \eta_2]$, and let $Y(\eta_2) = y(\eta_2)$. Then $Y(\eta) \geq y(\eta)$ for all $\eta_1 \leq \eta \leq \eta_2$. Setting $c = 1 - r$ in Lemma 8.5.2, we thus conclude from (8.5.16) that $Y(\eta_1) \to \infty$ for $\eta_1 \to \eta^*$ with a suitable number

$$\eta^* > \eta_2 \exp\left\{-2\pi/\sqrt{(1-r)(2n+1+r)}\right\}.$$

This means that the maximal interval of existence of (8.5.12) is contained in the finite interval $\left[\eta^*, \eta^* \exp\{2\pi/\sqrt{(1-r)(2n+1+r)}\}\right]$. Therefore (8.5.12) and consequently (8.5.3) is not solvable in $D(\gamma_2) = [\zeta_2, \infty)$. This proves assertion (II).

Example 7. Let $F = F_2$, $\delta_A = 0$ and $\gamma(\eta) = \gamma_1(\eta) = m(\zeta_1 - \eta)$. Then the following assertions hold:
(I) The constitutive equations (8.1.16), (8.1.17) belong to the class $T(\mathcal{MG})$, if and only if $D(\gamma_1) = [\zeta_2, \zeta_1)$ with $\zeta_2 \geq \zeta_1/2$.
(II) There exists $\zeta^* < \zeta_1/2$ such that for $D(\gamma_1) = [\zeta_2, \zeta_1]$ with $\zeta^* < \zeta_2 < \zeta_1/2$ the constitutive equations (8.1.16), (8.1.17) belong to TM.

These assertions should be compared to Example 3, where it is shown that the constitutive equations (8.1.16), (8.1.17) do not belong to $\mathcal{M}$ if $\zeta_2 < \zeta_1/2$, and that for $\zeta_2 = \zeta_1/2$ they belong to $\mathcal{M}$ if and only if $n \leq 15$. Assertions (I) and (II) together imply that $T(\mathcal{MG})$ is a proper subset of $TG \cap TM$, since for $F = F_2$ and $\zeta^* < \zeta_2 < \zeta_1/2$ the constitutive equations belong to TG and to TM, but not to $T(\mathcal{MG})$.

To prove (I), we use that

$$\frac{d}{d\eta}[\eta\gamma(\eta)] = \frac{d}{d\eta} m(\zeta_1\eta - \eta^2) = m(\zeta_1 - 2\eta) \begin{cases} \leq 0, \eta \geq \zeta_1/2 \\ > 0, \eta < \zeta_1/2, \end{cases}$$

which implies that (8.5.5) is satisfied if and only if $D(\gamma) = [\zeta_2, \zeta_1)$ with $\zeta_2 \geq \zeta_1/2$. Proposition 8.5.1 thus yields that the constitutive equations belong to $T(\mathcal{MG})$ if and only if $\zeta_2 \geq \zeta_1/2$.

To prove (II) we show that there exists $\zeta^* < \zeta_1/2$ such that for $\zeta^* < \zeta_2 < \zeta_1/2$ it is still possible to transform the constitutive equations to monotone type by choosing a function $\tilde{H}$ different from the one that transforms the constitutive equations to gradient type, such that (8.5.2) and (8.5.3) are satisfied in $[\zeta_2, \zeta_1)$. For example, for $a > 0$, $\theta = 1/n$ and $H_0 \in \mathbb{R}$ we might choose $\tilde{H}$ as solution of the following initial value problem:

$$\tilde{H}(\zeta_1/2) = H_0 \tag{8.5.18}$$

$$\tilde{H}'(\zeta) = \frac{1}{\sqrt{a(1 + \theta)\gamma(\zeta)\zeta}} \tag{8.5.19}$$

for $\zeta_1/2 \leq \zeta < \zeta_1$, and

$$\lim_{\zeta \nearrow \zeta_1/2} \tilde{H}(\zeta) = \tilde{H}(\zeta_1/2), \quad \lim_{\zeta \nearrow \zeta_1/2} \tilde{H}'(\zeta) = \tilde{H}'(\zeta_1/2) \tag{8.5.20}$$

$$Y'(\zeta) = -\frac{1}{4n\zeta}\left[\left((n+1)Y(\zeta) - \frac{n}{Y(\zeta)}\right)^2 Y(\zeta) + 2n\frac{(\zeta\gamma(\zeta))'}{\gamma(\zeta)}Y(\zeta)\right] \tag{8.5.21}$$

for $\zeta_2 \leq \zeta \leq \zeta_1/2$ with $Y(\zeta) = \sqrt{a\zeta\gamma(\zeta)}\,\tilde{H}'(\zeta)$. The initial value problem (8.5.18), (8.5.19) determines $\tilde{H}$ on the interval $[\zeta_1/2, \zeta_1)$, whereas the initial value problem (8.5.20), (8.5.21) determines $\tilde{H}$ on $[\zeta_2, \zeta_1/2]$. Note that (8.5.21) is a second order differential equation for $\tilde{H}$, which has a solution satisfying the initial conditions (8.5.20) if ζ_2 is chosen sufficiently close to $\zeta_1/2$. The initial conditions (8.5.20) are chosen such that $\tilde{H}$ is continuously differentiable on $[\zeta_2, \zeta_1)$.

(8.5.19) is chosen such that $\tilde{H}$ coincides on $[\zeta_1/2, \zeta_1)$ with the vector field that transforms the constitutive equations to gradient type. In the proof of statement (III) of Proposition 8.5.1 it is shown that (8.5.5) and (8.5.3) are equivalent if (8.5.19) holds. Since $\big(\zeta\gamma(\zeta)\big)' \leq 0$ on $[\zeta_1/2, \zeta_1)$, it thus follows that $\tilde{H}$ satisfies (8.5.3) on this interval. The relation (8.5.3) is also satisfied on the interval $[\zeta_2, \zeta_1/2)$, since (8.5.21) implies that the inequality (8.5.11) is satisfied, which is equivalent to (8.5.3). Therefore (8.5.3) is fulfilled on all of $[\zeta_2, \zeta_1)$, and from Proposition 8.5.1 we thus obtain that the constitutive equations (8.1.16), (8.1.17) belong to $\mathcal{TM}$ if the vector field $\tilde{H}$ also satisfies (8.5.2). Now, since $\tilde{H}'$ satisfies (8.5.19), it follows that the inequality (8.5.2) is satisfied on the interval $[\zeta_1/2, \zeta_1)$ if we choose H_0 in (8.5.18) not greater than $S(\zeta_1/2) = \sqrt{(1+\theta)\zeta_1/(2a\gamma(\zeta_1/2))}$. This is shown at the end of the proof of Proposition 8.5.1. By continuity, (8.5.2) is also satisfied on an interval $[\overline{\zeta}, \zeta_1/2]$ if we choose $H_0 = \tilde{H}(\zeta_1/2)$ small, and for sufficiently small H_0 it can even be achieved that $\overline{\zeta} = \zeta_2$, since the differential equation (8.5.21) determining $\tilde{H}$ on $[\zeta_2, \zeta_1/2)$ only contains the terms $\tilde{H}', \tilde{H}''$, but not $\tilde{H}$. This proves assertion (II).

Example 8. For $F = F_1 = d\exp(-\alpha s^{-n})$ and $\delta_A = 0$ the constitutive equations (8.1.16), (8.1.17) do not belong to $\mathcal{TM}$.

To see this, we use that F satisfies

$$F'(s) = \frac{\alpha n}{s^{n+1}} F(s).$$

If we insert this equation into (8.5.1) and set $s = |\tau|/\eta$ we obtain

$$\frac{4}{a} F^2 \left[\left(\frac{\alpha n}{s^n} \right)^2 \frac{\gamma}{\eta} - \frac{\alpha n}{s^n} \gamma' - \gamma \frac{\alpha n}{s^n} \frac{\tilde{H}''}{\tilde{H}'} \right] \geq F^2 \left[\frac{1}{a\eta\tilde{H}'} \frac{\alpha n}{s^n} + \left(\frac{\alpha n}{s^n} + 1 \right) \gamma \tilde{H}' \right]^2.$$

Here the function F depends on s, and the functions γ, $\tilde{H}$ depend on η. We use that $F(s) > 0$ for all $s > 0$ and collect terms with the same powers of s to get

$$0 \geq \left[\frac{1}{a\eta\tilde{H}'} - \gamma\tilde{H}' \right]^2 \left(\frac{\alpha n}{s^n} \right)^2$$

$$+ 2\left(\frac{\gamma}{a\eta} + (\gamma\tilde{H}')^2 + \frac{2}{a}\left(\gamma' + \gamma\frac{\tilde{H}''}{\tilde{H}'}\right) \right) \frac{\alpha n}{s^n} + (\gamma\tilde{H}')^2.$$

This inequality must hold for all $(\tau, \eta) = (P_0\sigma, \eta)$ with $(\sigma, \eta) \in \mathcal{S}^3 \times D(\gamma)$ and with $\tau \neq 0$, whence for all $(s, \eta) \in \mathbb{R}^+ \times D(\gamma)$. This is impossible, since for $s \to \infty$ the right hand side tends to the value $(\gamma(\eta)\tilde{H}'(\eta))^2 > 0$.

Chapter 9

Open Problems and Related Results

In this chapter we discuss possible generalizations and open problems of the theory developed in the preceding chapters and mention some research results published in recent years, mainly in the field of hysteresis operators, which are related to the topics of this work.

9.1 Transformation Fields Depending on ε and z

The investigations in Chapter 8 indicate that many of the examples of constitutive equations presented in Section 2.2 do not belong to the class $\mathcal{TM}$ and that this class is therefore to small. Existence and uniqueness of solutions of initial-boundary value problems to the system $(2.1.1) - (2.1.3)$ with constitutive equations from a much larger class could be obtained, if it would be possible to use transformation fields H which do not only depend on the vector z of internal variables, but also on the strain tensor $\varepsilon(\nabla_x u)$. This can be seen if we summarize the results of Chapters 5 and 6 as follows.

The transformed equations to the constitutive equations

$$T = D(\varepsilon - Bz) \tag{9.1.1}$$

$$z_t = f(\varepsilon, z) = g(\overline{L}\varepsilon - \hat{L}z) \tag{9.1.2}$$

are of monotone type if and only if a monotone vector field $g_H : \mathbb{R}^N \to \mathbb{R}^N$ and a suitable linear, symmetric, positive definite map $M : \mathbb{R}^N \to \mathbb{R}^N$ exist such that

$$T = D\big(\varepsilon - BH^{-1}(h)\big)$$

$$h_t = H'\big(H^{-1}(h)\big)\, g\big(\overline{L}\varepsilon - \hat{L}H^{-1}(h)\big) = g_H\big((BH^{-1})^T D\varepsilon - Mh\big) \tag{9.1.3}$$

holds. The function g_H can be determined from (9.1.3) by setting $\varepsilon = 0$:

$$g_H(h) = H'\big(H^{-1}(-M^{-1}h)\big)\, g\big(-\hat{L}H^{-1}(-M^{-1}h)\big). \tag{9.1.4}$$

Therefore (9.1.1), (9.1.2) are transformed to monotone type if and only if H and M satisfy the following three conditions:

(i) BH^{-1} must be linear.

(ii) g_H must be monotone. If $D(g_H)$ is convex, then g_H is convex if and only if the Jacobian $\nabla g_H(h)$ is positive semi-definite for every $h \in D(g_H)$. From (9.1.4) we see that this holds if and only if

$$\nabla_h \left[H'\left(H^{-1}(-M^{-1}h)\right) g\left(-\hat{L} H^{-1}\left(-M^{-1}h\right)\right)\right]$$

is positive semi-definite for every h.

(iii) Equation (9.1.3) must hold for all values of ε. If g_H is determined by (9.1.4), then (9.1.3) holds for $\varepsilon = 0$, but not necessarily for $\varepsilon \neq 0$. By Lemma 5.2.1, it also holds for $\varepsilon \neq 0$, if and only if $H'(z) g(\overline{L}\varepsilon - \hat{L}z)$ is constant on $V(c, M, H)$ for every c. One of the main results of Chapter 6 is that for many functions g this restricts the form of H to

$$H(z) = Pz + \tilde{H}(Qz).$$

It is this condition (iii) which is very restrictive, causing $\mathcal{TM}$ to be a relatively small class. This condition vanishes, if we allow the transformation field to depend not only on z, but also on $\varepsilon = \varepsilon(\nabla_x u)$. The new difficulty arising for such transformation fields is that after transformation the system of partial differential equations and constitutive equations is quasi-linear. If we denote the inverse of the mapping $z \mapsto H(\varepsilon, z)$ by H_ε^{-1}, then this system becomes

$$\rho u_{tt} = \mathrm{div}_x T$$

$$T = D(\varepsilon(\nabla_x u) - BH_\varepsilon^{-1}(h)) \tag{9.1.5}$$

$$h_t - [\nabla_\varepsilon H(\varepsilon, H_\varepsilon^{-1}(h))]\varepsilon_t = \nabla_z H(\varepsilon, H_\varepsilon^{-1}(h)) f(\varepsilon, H_\varepsilon^{-1}(h)), \tag{9.1.6}$$

which is quasilinear, because the term

$$\varepsilon_t = \varepsilon(\nabla_x u)_t = \frac{1}{2}\left(\nabla_x u_t + (\nabla_x u_t)^T\right)$$

consists of second order derivatives and the coefficient $\nabla_\varepsilon H(\varepsilon, H_\varepsilon^{-1}(h))$ depends on the solution. If it could be proved that initial-boundary value problems for these equations have solutions, provided that the pair

$$\left(\nabla_z H(\varepsilon, H_\varepsilon^{-1}(h)) f(\varepsilon, H_\varepsilon^{-1}(h)), \, BH_\varepsilon^{-1}(h)\right)$$

is of monotone type (in the sense of Definition 3.1.1, with BH_ε^{-1} linear and independent of ε), then a much larger class of transformation fields could be considered, and initial-boundary value problems with constitutive equations from a much larger class could be solved. This is an open problem.

9.2   History Functionals

Let $F\left([0, \infty), \mathbb{R}^n\right)$ denote the set of functions with domain $[0, \infty)$ and values in $\mathbb{R}^n$, and let R denote a suitable subset of $F\left([0, \infty), \mathbb{R}^n\right) \times \mathbb{R}^m$. A mapping

$$\mathop{\mathcal{F}}_{0 \leq \tau \leq t} : R \to F\left([0, \infty), \mathbb{R}^n\right)$$

is called history functional, if it has the property that to every $s \geq 0$ and to every pairs (ε_1, a), $(\varepsilon_2, a) \in R$ of functions $\varepsilon_i \in F([0, \infty), \mathbb{R}^n)$ and elements $a \in \mathbb{R}^m$ with $\varepsilon_1(\tau) = \varepsilon_2(\tau)$ for $0 \leq \tau \leq s$ the image functions $z_i = \mathop{\mathcal{F}}\limits_{0 \leq \tau \leq t}(\varepsilon_i, a)$ also satisfy $z_1(\tau) = z_2(\tau)$ for $0 \leq \tau \leq s$. History functionals are causal mappings. If

$$T = D(\varepsilon - Bz) \tag{9.2.1}$$

$$z_t \in f(\varepsilon, z) \tag{9.2.2}$$

are constitutive relations, which to every suitably given function $\varepsilon : [0, \infty) \to \mathcal{S}^3$ and initial data $z^{(0)} \in \mathbb{R}^N$ have a unique solution $(z, T) : [0, \infty) \to \mathbb{R}^N \times \mathcal{S}^3$ satisfying $z(0) = z^{(0)}$, then a history functional is defined by

$$\mathop{\mathcal{F}}\limits_{0 \leq \tau \leq t}(c, z^{(0)}) = T.$$

We say that the constitutive equations (9.2.1), (9.2.2) are a realization of the history functional $\mathop{\mathcal{F}}\limits_{0 \leq \tau \leq t}$. The transformed equations

$$T = D(\varepsilon - BH^{-1}(h))$$

$$h_t \in H'(H^{-1}(h)) f(\varepsilon, H^{-1}(h))$$

define the history functional $(\varepsilon, h^{(0)}) \mapsto \mathop{\mathcal{F}}\limits_{0 \leq \tau \leq t}(\varepsilon, H^{-1}(h^{(0)}))$, which differs from $\mathop{\mathcal{F}}\limits_{0 \leq \tau \leq t}$ only in the dependence on the initial data. Therefore we define these two history functionals to be equivalent. Consequently, all the constitutive equations which can be transformed into one another, form an equivalence class of realizations of equivalent history functionals. It is a fundamental open problem of the continuum mechanical theory of metals, to determine the history functionals which have realizations of the form (9.2.1), (9.2.2), and to develop a mathematical theory for the determination of a realization for a given history functional which is known from experimental measurements. This mathematical theory should replace the common procedure of formulating a set of constitutive equations, guided at least in part by intuitive ideas, and subsequently identifying the values of the parameters in these equations by comparison with experiments.

9.3 Hysteresis Operators. Existence Theory in L^p

A history functional, which is rate independent in the sense of Definition 2.1.1, is called a hysteresis operator. If the constitutive equations (9.2.1), (9.2.2) are rate independent, then they realize a hysteresis operators. Hysteresis operators are thus closely related to the problems studied in this work. In fact, since 1970, the larger part of the mathematical literature to initial-boundary value problems to constitutive models with internal variables concerns rate independent problems, and hence deals with hysteresis operators. As we outlined in Section 3.2, to prove existence of solutions for such rate independent initial-boundary value problems the approach using monotonicity properties of the operator defined by the constitutive relations in the Hilbert space L^2 is usually employed. However, for some initial-boundary value problems to partial differential equations combined with hysteresis operators, which differ from the initial-boundary value problems studied in this work, existence results have been published in recent years, whose proof is not based on this approach. Here we mention some of the ideas used in the proofs of these results and consider their relation to the problems studied in this work.

In his book on hysteresis problems [212, Chapt. VII], Visintin uses monotonicity in the Hilbert space L^2 to study some problems from elasto-plasticity. To study some other problems with hysteresis, however, in this book (Chapt. VIII) he uses the idea to employ the theory of nonlinear semigroups of accretive operators, which is a generalization of the Hilbert space theory of semi-groups of monotone operators to Banach spaces. Let X be a real Banach space with norm $\|\cdot\|$. An operator A : $D(A) \subseteq X \to \mathcal{P}(X)$ is said to be accretive if for all $u_1, u_2 \in D(A)$, all $w_1 \in A(u_1)$, $w_2 \in A(u_2)$, and all $\lambda > 0$, the inequality

$$\|u_1 - u_2\| \leq \|u_1 - u_2 + \lambda(w_1 - w_2)\| \tag{9.3.1}$$

holds. In a Hilbert space, this definition is equivalent to the ordinary definition of monotone operators used in Chapter 4. A theory of semigroups of accretive operators exists, cf. [65] for example, and Visintin applies this theory in the case of the Banach spaces L^1 and L^∞ to study several problems, where partial differential equations and hysteresis operators are coupled. It would be interesting to know whether this idea can also be used to study the initial-boundary value problem (2.1.1) – (2.1.9). For example, in the rate independent problems studied in Chapter 7 and with the notation of Corollary 7.2.3, one would expect that to obtain an accretive operator in a suitable L^1 space similar to the space used by Visintin, the mapping $g_k : \mathbb{R}^N \to \mathcal{P}(\mathbb{R}^N)$ must be accretive with respect to the norm

$$\|z\| = \|(\hat{z}, \tilde{z})\| = |\hat{z}| + |\tilde{z}|,$$

or a similar norm on $\mathbb{R}^N = R(\hat{L}^{-1}\overline{L}) \times \ker(B)$, where $|\cdot|$ denotes the ordinary Euclidean norm of $R(\hat{L}^{-1}\overline{L})$ and $\ker(B)$, respectively. It is not difficult to see that if g_K has the properties (7.1.5), (7.1.6), and if $z \in \partial K$, then this would imply that $g_K(z)$ need not be a ray normal to ∂K, as in the L^2 case discussed in Chapter 7, but would have to be a ray with a fixed "diagonal direction".

If existence results for initial-boundary value problems in such (special) L^1 and L^∞ spaces could be proved, one might also use transformation of interior variables to prove existence of solutions in L^1 and L^∞ for problems with constitutive equations, which originally do not yield accretive operators. These are open problems.

Another method used by Visintin in [212, Chapt. IX] to prove existence of solutions to partial differential equations with history functionals is based on the following condition for the history functional $\underset{0 \leq \tau \leq t}{\mathcal{F}}$. We formulate this condition in a way which is slightly different from the formulation in [212]. If $\varepsilon : [0, T] \to \mathbb{R}^n$ is affine in $[t_1, t_2]$, then the history functional must satisfy

$$\left([\underset{0 \leq \tau \leq t}{\mathcal{F}}(\varepsilon, a)](t_2) - [\underset{0 \leq \tau \leq t}{\mathcal{F}}(\varepsilon, a)](t_1) \right) \cdot \left(\varepsilon(t_2) - \varepsilon(t_1) \right) \geq 0. \tag{9.3.2}$$

This condition is well suited to hysteresis operators, but history functionals defined by rate dependent constitutive equations of the form

$$T = D(\varepsilon - Bz) \tag{9.3.3}$$

$$z_t = f(\varepsilon, z) \tag{9.3.4}$$

with f having suitable continuity properties never satisfy this condition (with the exception of the trivial case, where the history functional defined by (9.3.3), (9.3.4) is $\varepsilon \mapsto D\varepsilon$).

To see this, we note that if $\underset{0 \le \tau \le t}{\mathcal{F}}$ is defined by (9.3.3), (9.3.4), then (9.3.2) takes the form

$$[D(\varepsilon(t_2) - Bz(t_2)) - D(\varepsilon(t_1) - Bz(t_1))] \cdot (\varepsilon(t_2) - \varepsilon(t_1)) \ge 0,$$

from which we obtain

$$\left[\frac{d}{dt} D(\varepsilon(t_1) - Bz(t_1))\right] \cdot \frac{d}{dt} \varepsilon(t_1) \ge 0$$

for every continuously differentiable function ε and any solution z of (9.3.4) for this ε. Whence, using (9.3.4),

$$\left[D\big(\varepsilon_t(t_1) - Bz_t(t_1)\big)\right] \cdot \varepsilon_t(t_1) \qquad\qquad (9.3.5)$$

$$= \left[D\big(\varepsilon_t(t_1) - Bf(\varepsilon(t_1), z(t_1))\big)\right] \cdot \varepsilon_t(t_1) \ge 0.$$

Now choose a function ε^*, initial data $z^{(0)}$ and a time t_1 such that

$$w^* = Bf\big(\varepsilon^*(t_1), z^*(t_1)\big) \ne 0,$$

where z^* is the solution of $z_t^* = f(\varepsilon^*, z^*)$, $z^*(0) = z^{(0)}$. For $0 < \delta < t_1$ choose a function ε, which coincides with ε^* in the interval $[0, t_1 - \delta)$, and satisfies

$$\sup_{0 \le t \le t_1} |\varepsilon(t) - \varepsilon^*(t)| \le 1$$

$$\varepsilon_t(t_1) = \frac{1}{2} w^*.$$

For the solution z of

$$z_t = f(\varepsilon, z)$$

$$z(0) = z^{(0)}$$

it then follows from the usual theory of systems of ordinary differential equations that

$$Bf\big(\varepsilon(t_1), z(t_1)\big) \to w^*$$

as $\delta \to 0$. Hence

$$\left[D\big(\varepsilon_t(t_1) - Bf(\varepsilon(t_1), z(t_1))\big)\right] \cdot \varepsilon_t(t_1)$$

$$\to \left[D\big(\tfrac{1}{2} - 1\big)w^*\right] \cdot \left(\tfrac{1}{2} w^*\right) = -\frac{1}{4}(Dw^*) \cdot w^* < 0,$$

which implies that

$$\left[D\big(\varepsilon_t(t_1) - Bf(\varepsilon(t_1), z(t_1))\big)\right] \cdot \varepsilon_t(t_1) < 0$$

for all sufficiently small $\delta > 0$. Therefore (9.3.5) and consequently (9.3.2) is not satisfied for the rate dependent constitutive equations (9.3.3), (9.3.4).

This example illustrates that rate independence imposes a strong structure on history functionals, and that therefore many results which hold for rate independent problems cannot be expected to hold for rate dependent problems.

Another tool to prove existence for initial-boundary value problems with hysteresis operators are the energy estimates derived by Krejčí [121, 122]. With these estimates existence of solutions for initial-boundary value problems for partial differential equations with hysteresis operators having strictly convex hysteresis loops can be proved. For these estimates see also Visintin [212, Sect. IX.5] and the book of Brokate and Sprekels [25, Chapt. 3].

9.4   Constitutive Equations Defining Continuous Operators in Banach Spaces

A larger part of the book of Ionescu and Sofonea [106] is devoted to studying initial-boundary value problems for equations, which in our notation can be written as,

$$\rho u_{tt} = \operatorname{div}_x T + \rho b \tag{9.4.1}$$

$$T = D\big(\varepsilon(\nabla_x u) - \varepsilon_p\big) \tag{9.4.2}$$

$$z_t = f(x, t, T, \varepsilon(\nabla_x u), \kappa), \tag{9.4.3}$$

where $(x,t) \in \Omega \times [0,\infty)$, where $b(x,t)$ is a given function, and where the interior variables are $z(x,t) = \big(\varepsilon_p(x,t), \kappa(x,t)\big)$ with $\kappa(x,t) \in \mathbb{R}$. The function f is required to satisfy, besides some other technical conditions, the uniform Lipschitz condition

$$|f(x, t_1, T_1, \varepsilon_1, \kappa_1) - f(x, t_2, T_2, \varepsilon_2, \kappa_2)| \le$$
$$\le L\big(|\theta(x,t_1) - \theta(x,t_2)| + |T_1 - T_2| + |\varepsilon_1 - \varepsilon_2| + |\kappa_1 - \kappa_2|\big)$$

for a constant $L > 0$ and a function $\theta \in C([0,\infty), L^2(\Omega))$.

Because of this uniform Lipschitz condition, the nonlinear part in the equations (9.4.1) – (9.4.3) defines a continuous operator in suitable function spaces. This is used in [106] to prove existence, uniqueness, and continuous dependence of the solution employing functional analytic tools from the theory of continuous operators in Banach spaces.

The examples given in Section 2.2 generally do not satisfy this uniform Lipschitz condition, but it might be interesting to investigate whether transformation of interior variables can be used to transform constitutive equations not satisfying such a Lipschitz condition into equations for which it is satisfied, perhaps by applying singular transformation fields.

Appendix A

The Second Law of Thermodynamics and the Dissipation Inequality

In this appendix we show how the dissipation inequality (2.1.9) is derived from the second law of thermodynamics. Our restricted goal is to give a short derivation with an introductory character for readers not familiar with thermodynamics. Therefore at several instances our assumptions are simpler than the most general ones under which the dissipation inequality can be derived. Also, we cannot discuss problems like definition of the temperature, the choice of the entropy flux, the choice of the material class, and other advanced problems, and we do not use the general method introduced by Liu and Müller [136, 139] to exploit the second law of thermodynamics to derive restrictions for the constitutive equations and the dissipation inequality. For the special and relatively simple system of equations considered below, a direct approach leads to the dissipation inequality in a shorter way. For all these questions the reader is refered to the literature in thermodynamics, for example [8, 48, 94, 95, 96, 103, 105, 159, 211].

A.1 Consequences of the Second Law for the Constitutive Equations

In the preceding chapters we neglected the dependence of the constitutive equations on the temperature distribution in the solid body. However, to derive the dissipation inequality (2.1.9), this dependence must be taken into account. Now, if the constitutive equations depend on the temperature, then, of course, in order to compute the deformation of a solid body and the stress distribution in this body, the temperature distribution must also be determined. Here we start with the formulation of the system of differential equations from which these unknown fields can be determined. Expositions of the derivation of these differential equations can be found in [47, 94, 95, 96], for example.

Let $\Omega \subseteq \mathbb{R}^3$ be an open set with smooth boundary. Ω represents a solid body at time $t = 0$ and is called the reference configuration. For $x \in \Omega$ and $t \in \mathbb{R}$, let $\varphi(x, t) \in \mathbb{R}^3$ be the position of the material point at time t which at time $t = 0$ is situated at x. This implies that $x \mapsto \varphi(x, 0) : \Omega \to \Omega$ is the identity. The 3×3–matrix $\nabla_x \varphi(x, t)$ of first order partial derivatives with respect to the components of x is called the deformation gradient. Since two different material points cannot be located at the same position, and since the deformation of the body should be sufficiently regular, one requires that the mapping $x \mapsto \varphi(x, t) : \Omega \to \mathbb{R}^3$ is injective for all t and that the matrix $\nabla_x \varphi(x, t)$ is non-singular for all (x, t). Together with $\nabla_x \varphi(x, 0) = I$, this implies $\det(\nabla_x \varphi(x, t)) > 0$ for all (x, t).

We consider simple materials, which means that the value of the stress tensor at the point (x, t) only depends on the values of the density, the temperature, the internal variables and the deformation gradient at the same point. However, the stress tensor does not directly depend on $\nabla_x \varphi(x, t)$, but on the matrix

$$E(x, t) = \frac{1}{2}\Big([\nabla_x \varphi(x, t)]^T \nabla_x \varphi(x, t) - I\Big) \in \mathbb{R}^3,$$

the Green strain tensor. This is a consequence of the principle of material objectivity, cf. [47, 94, 95, 159] for example. Since $\nabla_x \varphi(x, t)$ is non-singular,

$$2E(x, t) + I = \nabla_x \varphi(x, t)^T \nabla_x \varphi(x, t)$$

is a symmetric, positive definite matrix. $z(x, t) \in \mathbb{R}^N$ denotes the vector of internal variables. Instead of the temperature, one first introduces the internal energy $e(x, t)$ as an unknown. The temperature θ is then a function of (e, E, z) :

$$\theta(x, t) = \hat{\theta}(e(x, t), E(x, t), z(x, t)).$$

As usual, in the following we do not distinguish in notation between the two different functions θ and $\hat{\theta}$ on both sides of this equation and write θ for both functions. With the exception of the stress tensor we use this convention also for the other thermodynamic variables.

With the density $\rho(x, t)$, the deformation velocity

$$v(x, t) = \varphi_t(x, t),$$

the first Piola-Kirchhoff stress tensor $T_R(x, t)$, which is a 3×3–matrix, the second Piola-Kirchhoff stress tensor

$$\tilde{T}(x, t) = [\nabla_x \varphi(x, t)]^{-1} T_R(x, t)$$

and the heat flow $q(x, t) \in \mathbb{R}^3$, the laws of conservation of mass, momentum, energy and the evolution equation for the internal variables read in Lagrangian coordinates

$$\rho_t = 0 \tag{A.1.1}$$

$$\rho v_t = \mathrm{div}_x T_R \tag{A.1.2}$$

$$\rho e_t = \tilde{T} \cdot E_t - \mathrm{div}_x q \tag{A.1.3}$$

$$z_t = f. \tag{A.1.4}$$

In addition, one needs the constitutive equations

$$\tilde{T} = \hat{T}(\rho, e, E, z) \tag{A.1.5}$$

$$q = q(e, E, z, \nabla_x e, \nabla_x E, \nabla_x z) \tag{A.1.6}$$

$$f = f(e, E, z). \tag{A.1.7}$$

The system is closed if one adds the equations defining E, v, $\tilde{T}$:

$$E = \frac{1}{2}[(\nabla_x\varphi)^T(\nabla_x\varphi) - I] \qquad \text{(A.1.8)}$$

$$v = \varphi_t \qquad \text{(A.1.9)}$$

$$T_R = (\nabla_x\varphi)\,\tilde{T}. \qquad \text{(A.1.10)}$$

As a consequence of the law of conservation of the moment of momentum, the second Piola-Kirchhoff stress tensor must be a symmetric matrix, hence

$$\tilde{T}(x, t) \in \mathcal{S}^3$$

for every (x, t). This must be observed in the formulation of the constitutive equation (A.1.5). The initial conditions for the solution (ρ, φ, e, z) of this system at the time t_0 are

$$\begin{aligned}
\rho(x, t_0) = \rho^{(0)}(x), \quad \varphi(x, t_0) = \varphi^{(0)}(x), \quad \varphi_t(x, t_0) = v^{(0)}(x), \\
e(x, t_0) = e^{(0)}(x), \quad z(x, t_0) = z^{(0)}(x).
\end{aligned} \qquad \text{(A.1.11)}$$

One also needs boundary conditions, but since they play no role in the following considerations, we do not state them.

Under suitable assumptions for the constitutive equations and the initial data, a unique solution of this initial-boundary value problem exists locally in time. The second law of thermodynamics requires that for this initial-boundary value problem a function $s = s(e, E, z)$, the entropy, exists, such that for any solution (ρ, φ, e, z) of the problem representing the evolution of the deformation and energy distribution of the solid body the function

$$s(x, t) = s(e(x, t), E(x, t), z(x, t)) \qquad \text{(A.1.12)}$$

satisfies the Clausius-Duhem inequality

$$\rho s_t + \operatorname{div}_x\left(\frac{q}{\theta}\right) \geq 0 \qquad \text{(A.1.13)}$$

for all (x, t) in the domain of the solution. It will be shown that if the set of solutions satisfying (A.1.13) contains all solutions to a relatively small class of initial data specified below, then this is only possible if the constitutive equations and the functions $s(e, E, z)$, $\theta(e, E, z)$ satisfy some restrictive conditions, which in particular imply that the dissipation inequality (2.1.9) must be satisfied.

To this end we introduce some notations and specify the requirements and assumptions for the given functions $\hat{T}$, q, f, s and θ more precisely. Since $E(x, t) \in \mathcal{S}^3$ and since $2E(x, t) + I$ must be positive definite, we assume that the common domain of definition of f, s and θ is the set $\mathbb{R} \times \hat{\mathcal{S}}^3 \times \mathcal{D}$, where

$$\hat{\mathcal{S}}^3 = \{S \in \mathcal{S}^3 \mid 2S + I \text{ is positive definite}\}$$

and where $\mathcal{D}$ is an open subset of $\mathbb{R}^N$, hence $s : \mathbb{R} \times \hat{\mathcal{S}}^3 \times \mathcal{D} \to \mathbb{R}$, $\theta : \mathbb{R} \times \hat{\mathcal{S}}^3 \times \mathcal{D} \to \mathbb{R}^+$ and $f : \mathbb{R} \times \hat{\mathcal{S}}^3 \times \mathcal{D} \to \mathbb{R}^N$. We assume that the domain of definition of $\hat{T}$ is $\mathbb{R}_0^+ \times \mathbb{R} \times \hat{\mathcal{S}}^3 \times \mathcal{D}$, hence $\hat{T} : \mathbb{R}_0^+ \times \mathbb{R} \times \hat{\mathcal{S}}^3 \times \mathcal{D} \to \mathcal{S}^3$.

Since $x \mapsto E(x, t) : \Omega \subseteq \mathbb{R}^3 \to \mathcal{S}^3$, for every (x, t) the derivative $\nabla_x E(x, t)$ is a linear mapping from $\mathbb{R}^3$ to $\mathcal{S}^3$ and can be represented by an element of $(\mathcal{S}^3)^3$, whereas $\nabla_x z(x, t)$ is a linear mapping from $\mathbb{R}^3$ to $\mathbb{R}^N$ and can be represented by an element of $(\mathbb{R}^N)^3$. Since $\nabla_x e(x, t) \in \mathbb{R}^3$, we thus choose for the domain $D(q)$ of $q : D(q) \to \mathbb{R}^3$ the set $\mathbb{R} \times \hat{\mathcal{S}}^3 \times \mathcal{D} \times \mathbb{R}^3 \times (\mathcal{S}^3)^3 \times (\mathbb{R}^N)^3$. We assume that $\hat{T}$, q and f are continuously differentiable functions.

The above choice of the domain of s implies that $\nabla_E s(e, E, z)$ is a linear mapping from $\mathcal{S}^3$ to $\mathbb{R}$ and can be represented by a symmetric matrix, whence $\nabla_E s(e, E, z) \in \mathcal{S}^3$.

Finally, by $\mathcal{C}$ we denote the class of all functions

$$(\rho^{(0)}, \varphi^{(0)}, v^{(0)}, e^{(0)}, z^{(0)}) \in C^\infty(\Omega, \mathbb{R}^+ \times \mathbb{R}^3 \times \mathbb{R}^3 \times \mathbb{R} \times \mathcal{D}),$$

for which $(\rho^{(0)}, \nabla\varphi^{(0)}, v^{(0)}, e^{(0)}, z^{(0)})$ is constant outside a compact set, and which satisfy $\det(\nabla\varphi^{(0)}(x)) > 0$ for all $x \in \Omega$.

Theorem A.1.1 *Assume that for all initial data $(\rho^{(0)}, \varphi^{(0)}, v^{(0)}, e^{(0)}, z^{(0)})$ from the class $\mathcal{C}$, the initial-boundary value problem to (A.1.1) – (A.1.11) has a solution, which exists locally in time and is three times continuously differentiable. Moreover, assume that*

$$\nabla_A q(a, A) \neq 0 \tag{A.1.14}$$

for all $a \in \mathbb{R} \times \hat{\mathcal{S}}^3 \times \mathcal{D}$ and all $A \in \mathbb{R}^3 \times (\mathcal{S}^3)^3 \times (\mathbb{R}^N)^3$. Let $s \in C^1(\mathbb{R} \times \hat{\mathcal{S}}^3 \times \mathcal{D}, \mathbb{R})$ and $\theta \in C^1(\mathbb{R} \times \hat{\mathcal{S}}^3 \times \mathcal{D}, \mathbb{R}^+)$ be given. Then the following statements (i) and (ii) are equivalent:

(i)   *For every three times continuously differentiable solution (ρ, φ, e, z) of the equations (A.1.1) – (A.1.10) the functions $(x, t) \mapsto s(e(x, t), E(x, t), z(x, t))$ and $(x, t) \mapsto \theta(e(x, t), E(x, t), z(x, t))$ satisfy the Clausius-Duhem inequality (A.1.13) in the domain of definition of this solution.*

(ii)   *For every $(\rho^*, a, A) = (\rho^*, a_1, a_2, a_3, A_1, A_2, A_3) \in \mathbb{R}^+ \times (\mathbb{R} \times \hat{\mathcal{S}}^3 \times \mathcal{D}) \times (\mathbb{R}^3 \times (\mathcal{S}^3)^3 \times (\mathbb{R}^N)^3)$, the functions s, $\hat{T}$, θ, q and f satisfy the four relations*

$$\hat{T}(\rho^*, a) = -\rho^* \left[\frac{\partial s}{\partial e}(a)\right]^{-1} \nabla_E s(a) \tag{A.1.15}$$

$$\theta(a) = \left[\frac{\partial s}{\partial e}(a)\right]^{-1} = \frac{\partial}{\partial s} e(s, a_2, a_3)\big|_{s=s(a)} \tag{A.1.16}$$

$$\nabla_z s(a) \cdot f(a) \geq 0 \tag{A.1.17}$$

$$q(a, A) \cdot \left[\frac{\partial \theta}{\partial e}(a) A_1 + (\nabla_E \theta(a) A_2)^T + (\nabla_z \theta(a) A_3)^T\right] \leq 0. \tag{A.1.18}$$

In the second equation, $s \mapsto e(s, E, z)$ is the inverse of the function $e \mapsto s(e, E, z)$. In the last equation, $\nabla_E \theta(a) A_2 \in \mathbb{R}^3$ denotes the matrix representation of the linear mapping from $\mathbb{R}^3$ to $\mathbb{R}$ obtained by composition of the linear mapping $\nabla_E \theta(a) : \mathcal{S}^3 \to \mathbb{R}$ with $A_2 : \mathbb{R}^3 \to \mathcal{S}^3$, and $\nabla_z \theta(a) A_3 \in \mathbb{R}^3$ denotes the matrix representation of the linear mapping from $\mathbb{R}^3$ to $\mathbb{R}$ obtained by composition of $\nabla_z \theta(a) : \mathbb{R}^N \to \mathbb{R}$ with $A_3 : \mathbb{R}^3 \to \mathbb{R}^N$.

Proof. Let $(\rho, \varphi, e, z) = (\rho(x,t), \varphi(x,t), e(x,t), z(x,t))$ be a solution of (A.1.1) – (A.1.10). Insertion of e, E, z, $\nabla_x e$, $\nabla_x E$, $\nabla_x z$ into the arguments of s, q and θ and application of the chain rule of differentiation and of (A.1.3), (A.1.4), (A.1.5) yields

$$\rho s_t + \mathrm{div}_x \left(\frac{q}{\theta}\right) = \rho \left(\frac{\partial s}{\partial e} e_t + \nabla_E s \cdot E_t + \nabla_z s \cdot z_t\right) + \mathrm{div}_x \left(\frac{q}{\theta}\right) \qquad (A.1.19)$$

$$= \frac{\partial s}{\partial e}(\hat{T} \cdot E_t - \mathrm{div}_x q) + \rho \nabla_E s \cdot E_t + \rho \nabla_z s \cdot z_t + \frac{1}{\theta}\mathrm{div}_x q - \frac{1}{\theta^2} q \cdot \nabla_x \theta$$

$$= \left(\frac{\partial s}{\partial e}\hat{T} + \rho \nabla_E s\right) \cdot E_t - \left(\frac{\partial s}{\partial e} - \frac{1}{\theta}\right) \mathrm{div}_x q + \rho \nabla_z s \cdot f - \frac{1}{\theta^2} q \cdot \nabla_x \theta .$$

The Clausius-Duhem inequality (A.1.13) is satisfied if and only if the left hand side and therefore the right hand side of this equality is non-negative at every point (x,t) in the domain of (ρ, φ, e, z). Since

$$\nabla_x \theta = \frac{\partial \theta}{\partial e}\nabla_x e + (\nabla_E \theta \, \nabla_x E)^T + (\nabla_z \theta \, \nabla_x z)^T \qquad (A.1.20)$$

with $(\nabla_x e(x,t), \nabla_x E(x,t), \nabla_x z(x,t)) \in \mathbb{R}^3 \times (\mathcal{S}^3)^3 \times (\mathbb{R}^N)^3$, the relations (A.1.15) – (A.1.18) imply that the right hand side of (A.1.19) is non-negative, and thus that the Clausius-Duhem inequality is satisfied for every three times continuously differentiable solution of (A.1.1) – (A.1.10) in the domain of this solution. It remains to prove the converse.

To this end we fix $x_0 \in \Omega$ and assume that (A.1.13) holds for every solution for initial data from the class $\mathcal{C}$. This implies in particular that the right hand side of (A.1.19) must be non-negative at the point (x_0, t_0) for all such solutions (ρ, φ, e, z). In the following we show that to every $\rho^* \in \mathbb{R}^+$, $a \in \mathbb{R} \times \hat{\mathcal{S}}^3 \times D$, $A \in \mathbb{R}^3 \times (\mathcal{S}^3)^3 \times (\mathbb{R}^N)^3$, $r \in \mathbb{R}$ and $S \in \mathcal{S}^3$ we can find initial data $(\rho^{(0)}, \varphi^{(0)}, v^{(0)}, e^{(0)}, z^{(0)}) \in \mathcal{C}$ with

$$\rho(x_0, t_0) = \rho^* \qquad (A.1.21)$$

$$(e(x_0, t_0), E(x_0, t_0), z(x_0, t_0)) = a \qquad (A.1.22)$$

$$(\nabla_x e(x_0, t_0), \nabla_x E(x_0, t_0), \nabla_x z(x_0, t_0)) = A \qquad (A.1.23)$$

$$E_t(x_0, t_0) = S \qquad (A.1.24)$$

$$\mathrm{div}_x q(x_0, t_0) = r . \qquad (A.1.25)$$

We now evaluate (A.1.19) for solutions to these initial data at (x_0, t_0). The following three expressions, which appear in (A.1.19), only depend on (ρ^*, a, A) and are independent of r and S, since

$$\frac{\partial s}{\partial e}\hat{T} + \rho \nabla_E s = \frac{\partial s}{\partial e}(a)\hat{T}(\rho^*, a) + \rho^* \nabla_E s(a) \in \mathcal{S}^3$$

$$\frac{\partial s}{\partial e} - \frac{1}{\theta} = \frac{\partial s}{\partial e}(a) - \frac{1}{\theta(a)} \in \mathbb{R}$$

$$\rho \nabla_z s \cdot f - \frac{1}{\theta^2} q \cdot \nabla_x \theta = \rho^* \nabla_z s(a) \cdot f(a)$$

$$- \frac{1}{\theta(a)^2} q(a, A) \cdot \left[\frac{\partial \theta}{\partial e}(a)A_1 + (\nabla_E \theta(a) A_2)^T + (\nabla_z \theta(a) A_3)^T\right] ,$$

where in the derivation of the last equation we used (A.1.20). Therefore the values of these expressions do not change if we fix ρ^*, a, and A, and vary $r = \text{div}_x q(x_0, t_0)$ and $S = E_t(x_0, t_0)$ arbitrarily. This implies that the right hand side of (A.1.19) can be non-negative for all solutions to initial data from the class $\mathcal{C}$ only if (A.1.15), (A.1.16) hold and if

$$\rho^* \nabla_z s(a) \cdot f(a) - \frac{1}{\theta(a)^2} g(a, A) \cdot \left[\frac{\partial \theta}{\partial e}(a) A_1 + (\nabla_E \theta(a) A_2)^T + (\nabla_z \theta(a) A_3)^T \right] \geq 0.$$

From this inequality we obtain (A.1.17) if we choose $A = 0$, while (A.1.18) is obtained by fixing a and A and sending $\rho^* \to 0$.

Thus, to complete the proof it must be shown that (A.1.21) – (A.1.25) can be satisfied. Let $\rho^* > 0$, $a = (a_1, a_2, a_3) \in \mathbb{R} \times \hat{\mathcal{S}}^3 \times \mathcal{D}$, $A = (A_1, A_2, A_3) \in \mathbb{R}^3 \times (\mathcal{S}^3)^3 \times (\mathbb{R}^N)^3$, $S \in \mathcal{S}^3$ and $r \in \mathbb{R}$ be given. We first note that (A.1.8) implies

$$E_{x_i}(x_0, t_0) = \frac{1}{2} \left[(\nabla \varphi_{x_i}^{(0)})^T \nabla \varphi^{(0)} + (\nabla \varphi^{(0)})^T \nabla \varphi_{x_i}^{(0)} \right] \tag{A.1.26}$$

and

$$\nabla E_{x_i}(x_0, t_0) = (E_{x_i x_1}, E_{x_i x_2}, E_{x_i x_3})$$
$$= \frac{1}{2} \Big((\nabla \varphi_{x_i x_j}^{(0)})^T \nabla \varphi^{(0)} + (\nabla \varphi^{(0)})^T \nabla \varphi_{x_i x_j}^{(0)} \tag{A.1.27}$$
$$+ (\nabla \varphi_{x_j}^{(0)})^T \nabla \varphi_{x_i}^{(0)} + (\nabla \varphi_{x_i}^{(0)})^T \nabla \varphi_{x_j}^{(0)} \Big)_{j=1,2,3},$$

whereas (A.1.8) and (A.1.11) yield

$$E_t(x_0, t_0) = \frac{1}{2} \left[(\nabla \varphi^{(0)})^T \nabla_x \varphi_t + (\nabla_x \varphi_t)^T \nabla \varphi^{(0)} \right] \tag{A.1.28}$$
$$= \frac{1}{2} \left[(\nabla \varphi^{(0)})^T \nabla v^{(0)} + (\nabla v^{(0)})^T \nabla \varphi^{(0)} \right].$$

Moreover, with $(b, B) = (b_1, b_2, b_3, B_1, B_2, B_3) = (e, E, z, \nabla_x e, \nabla_x E, \nabla_x z)(x_0, t_0)$ and with $q(b, B) = (q_1(b, B), q_2(b, B), q_3(b, B))$ the chain rule implies that

$$\text{div}_x q(x_0, t_0) = \sum_{i=1}^{3} \left[\frac{\partial q_i}{\partial b_1}(b, B) e_{x_i}^{(0)}(x_0) + (\nabla_{b_2} q_i(b, B)) \cdot E_{x_i}(x_0, t_0) \right.$$
$$+ (\nabla_{b_3} q_i(b, B)) \cdot z_{x_i}^{(0)}(x_0) + (\nabla_{B_1} q_i(b, B)) \cdot \nabla e_{x_i}^{(0)}(x_0) \tag{A.1.29}$$
$$+ (\nabla_{B_2} q_i(b, B)) \cdot \nabla E_{x_i}(x_0, t_0) + (\nabla_{B_3} q_i(b, B)) \cdot \nabla z_{x_i}^{(0)}(x_0) \Big].$$

Since for $a_2 \in \hat{\mathcal{S}}^3$ the matrix $2a_2 + I$ is symmetric and positive definite, we can select a 3×3–matrix M with $\det M > 0$ and with

$$a_2 = \frac{1}{2}[M^T M - I], \tag{A.1.30}$$

for example by choosing $M = \sqrt{2a_2 + I}$. Because M is invertible, to $A_2 = (A_2^{(1)}, A_2^{(2)}, A_2^{(3)}) \in (S^3)^3$ and $S \in S^3$ we can then select 3×3-matrices $M^{(i)}$, V with

$$A_2^{(i)} = \frac{1}{2}[M^{(i)T}M + M^T M^{(i)}] \tag{A.1.31}$$

$$S = \frac{1}{2}[M^T V + V^T M]. \tag{A.1.32}$$

Moreover, using the assumption (A.1.14), to $r \in \mathbb{R}$ we can select vectors $m^{(i)} \in \mathbb{R}^3$ and matrices $Z^{(i)} \in (\mathbb{R}^N)^3$, $B^{(i)} = (B^{(i1)}, B^{(i2)}, B^{(i3)}) \in (S^3)^3$, such that with $A_1 = (A_1^{(1)}, A_1^{(2)}, A_1^{(3)}) \in \mathbb{R}^3$ and $A_3 = (A_3^{(1)}, A_3^{(2)}, A_3^{(3)}) \in (\mathbb{R}^N)^3$

$$r = \sum_{i=1}^{0} \left[\frac{\partial q_i}{\partial a_1}(a, A)A_1^{(i)} + \nabla_{a_2}\, q_i(a, A) \cdot A_2^{(i)} \right.$$
$$+ \nabla_{a_3}\, q_i(a, A) \cdot A_3^{(i)} + \nabla_{A_1}\, q_i(a, A) \cdot m^{(i)} \tag{A.1.33}$$
$$\left. + \nabla_{A_2}\, q_i(a, A) \cdot B^{(i)} + \nabla_{A_3}\, q_i(a, A) \cdot Z^{(i)} \right].$$

Finally, using again that M is invertible, we can choose 3×3-matrices $M^{(ij)}$, $i, j = 1, 2, 3$, such that

$$B^{(ij)} = \frac{1}{2}[M^{(ij)T}M + M^T M^{(ij)} + M^{(j)T}M^{(i)} + M^{(i)T}M^{(j)}]. \tag{A.1.34}$$

Comparing now (A.1.30) with (A.1.8), (A.1.31) with (A.1.26), (A.1.32) with (A.1.28), (A.1.34) with (A.1.27) and (A.1.33) with (A.1.29), we see that the equations (A.1.21) – (A.1.25) are satisfied, if a function $(\rho^{(0)}, \varphi^{(0)}, v^{(0)}, e^{(0)}, z^{(0)}) \in C$ exists with

$$\rho^{(0)}(x_0) = \rho^*$$
$$e^{(0)}(x_0) = a_1, \quad \nabla e^{(0)}(x_0) = A_1, \quad \nabla e_{x_i}^{(0)}(x_0) = m^{(i)}$$
$$\nabla \varphi^{(0)}(x_0) = M, \quad \nabla \varphi_{x_i}^{(0)}(x_0) = M^{(i)}, \quad \nabla \varphi_{x_i x_j}^{(0)}(x_0) = M^{(ij)}$$
$$\nabla v^{(0)}(x_0) = V$$
$$z^{(0)}(x_0) = a_3, \quad \nabla z^{(0)}(x_0) = A_3, \quad \nabla z_{x_i}^{(0)}(x_0) = Z^{(i)},$$

where $i, j = 1, 2, 3$. However, such a function can be constructed by a standard and well known method. The proof of Theorem A.1.1 is complete.

A.2 The Free Energy

As is seen from (A.1.20), condition (A.1.18) implies for a solution of the system (A.1.1) – (A.1.10) that

$$q \cdot \nabla_x \theta \le 0,$$

which means that the heat flow q is directed from regions of higher temperature to regions of lower temperature. For such a solution (ρ, φ, e, z) conditions (A.1.15) – (A.1.17) together with (A.1.5) imply

$$\tilde{T} = -\rho \left[\frac{\partial s}{\partial e}(e, E, z) \right]^{-1} \nabla_E s(e, E, z) \tag{A.2.1}$$

$$\theta = \left[\frac{\partial s}{\partial e}(e, E, z) \right]^{-1} = \frac{\partial}{\partial s} e(s, E, z)|_{s=s(e,E,z)} \tag{A.2.2}$$

$$\nabla_z s(e, E, z) \cdot f(e, E, z) \geq 0. \tag{A.2.3}$$

The last relation is the inequality of internal dissipation. We show in several steps that it leads to the dissipation inequality (2.1.9).

In a first step we simplify these relations slightly by writing them all in a form with (s, E, z) as independent variables. To this end we differentiate the relation

$$s(e(s, E, z), E, z) = s$$

with respect to E and obtain

$$\frac{\partial s}{\partial e}(e(s, E, z), E, z) \nabla_E e(s, E, z) + (\nabla_E s)(e(s, E, z), E, z) = 0,$$

hence

$$(\nabla_E e)(s(e, E, z), E, z) = - \left[\frac{\partial s}{\partial e}(e, E, z) \right]^{-1} \nabla_E s(e, E, z).$$

In the same way we conclude that

$$(\nabla_z e)(s(e, E, z), E, z) = - \left[\frac{\partial s}{\partial e}(e, E, z) \right]^{-1} \nabla_z s(e, E, z)$$

$$= -\theta(e, E, z) \nabla_z s(e, E, z),$$

where in the second step we used (A.2.2). Since by assumption $\theta > 0$, insertion of these equations into (A.2.1) – (A.2.3) results in

$$\tilde{T} = \rho \nabla_E e(s, E, z) \tag{A.2.4}$$

$$\theta = \frac{\partial}{\partial s} e(s, E, z) \tag{A.2.5}$$

$$\nabla_z e(s, E, z) \cdot f(s, E, z) \leq 0, \tag{A.2.6}$$

where, by a slight abuse of notation, $f(s, E, z)$ is used to denote the function $f(e(s, E, z), E, z)$. Equations (A.2.4) and (A.2.5) imply that $(s, E) \mapsto e(s, E, z)$ is a potential for the field $(\theta, \frac{1}{\rho}\tilde{T})$.

Since the temperature θ can be measured easily, we next transform (A.2.4) – (A.2.6) in a form with θ as independent variable instead of s. For this purpose we note that because of $\theta(s, E, z) = (\partial/\partial s)e(s, E, z)$, the Legendre transform of the function $s \mapsto e(s, E, z)$ yields a function, which depends on (θ, E, z) and is a potential for the function $s(\theta, E, z)$.

More precisely, if one defines a function $L = (L_1, L_2, L_3) : \mathbb{R}^3 \to \mathbb{R}^3$ by

$$L_1(\xi, h, p) = p$$
$$L_2(\xi, h, p) = \xi p - h$$
$$L_3(\xi, h, p) = \xi,$$

and if $\xi \mapsto g(\xi)$ is a strictly convex function, then the equation

$$\hat{g}(L_1(\xi, g(\xi), g'(\xi))) = L_2(\xi, g(\xi), g'(\xi))$$

and whence

$$\hat{g}(g'(\xi)) = \xi g'(\xi) - g(\xi), \tag{A.2.7}$$

defines a convex function $\hat{g}$, the Legendre transform of g, satisfying

$$\hat{g}'(L_1(\xi, g(\xi), g'(\xi))) = L_3(\xi, g(\xi), g'(\xi)),$$

hence

$$\hat{g}'(g'(\xi)) = \xi . \tag{A.2.8}$$

Thus, if $s \mapsto e(s, E, z)$ is strictly convex for all (E, z), we can compute the Legendre transform of this function, which we denote by $-\psi$. Since the temperature satisfies $\theta = (\partial/\partial s)e$, we obtain from (A.2.7) that ψ is defined by

$$\psi(\theta(s, E, z), E, z) = e(s, E, z) - s\theta(s, E, z) \tag{A.2.9}$$

and, as a consequence of (A.2.8), satisfies

$$\frac{\partial}{\partial \theta}\psi(\theta(s, E, z), E, z) = -s, \tag{A.2.10}$$

hence

$$\frac{\partial}{\partial \theta}\psi(\theta, E, z) = -s(\theta, E, z),$$

with the inverse $\theta \mapsto s(\theta, E, z)$ of $s \mapsto \theta(s, E, z)$. The function ψ is called the (Helmholtz) free energy.

Equation (A.2.10) is the analogue of (A.2.5), but ψ also satisfies relations analogue to (A.2.4) and (A.2.6). To see this, we differentiate (A.2.9) with respect to E and to z, respectively, and obtain

$$\frac{\partial \psi}{\partial \theta}\nabla_E \theta + \nabla_E \psi = \nabla_E e - s\nabla_E \theta$$
$$\frac{\partial \psi}{\partial \theta}\nabla_z \theta + \nabla_z \psi = \nabla_z e - s\nabla_z \theta .$$

Application of (A.2.10) yields

$$\nabla_E \psi(\theta, E, z)\big|_{\theta=\theta(s,E,z)} = \nabla_E e(s, E, z)$$
$$\nabla_z \psi(\theta, E, z)\big|_{\theta=\theta(s,E,z)} = \nabla_z e(s, E, z),$$

which together with (A.2.4), (A.2.6) and (A.2.10) implies the following result:

Theorem A.2.1 *Let* (A.1.15) – (A.1.18) *be satisfied. If the function* $s \mapsto e(s, E, z)$ *is strictly convex for all* (E, z), *then the free energy* ψ, *defined as the negative of the Legendre transform of* $s \mapsto e(s, E, z)$, *satisfies the relations*

$$\tilde{T} = \rho \nabla_E \psi(\theta, E, z) \tag{A.2.11}$$

$$s = -\frac{\partial}{\partial \theta} \psi(\theta, E, z)$$

$$\nabla_z \psi(\theta, E, z) \cdot f(\theta, E, z) \leq 0. \tag{A.2.12}$$

Equation (2.1.8) and the dissipation inequality (2.1.9) are obtained from these relations by assuming that in the problems studied the temperature dependence of ψ and f can be neglected and that only small deformations occur, which allows us to linearize the constitutive equations in part. For this linearization one introduces the displacement

$$u(x, t) = \varphi(x, t) - x.$$

Then $\nabla_x \varphi(x, t) = \nabla_x u(x, t) + I$, whence

$$E = \frac{1}{2}[(\nabla_x \varphi)^T \nabla_x \varphi - I] = \frac{1}{2}[\nabla_x u + (\nabla_x u)^T + (\nabla_x u)^T \nabla_x u]$$

$$= \varepsilon(\nabla_x u) + \frac{1}{2}(\nabla_x u)^T \nabla_x u.$$

This implies that

$$E = \varepsilon(\nabla_x u) + O(|\nabla_x u|^2), \quad |\nabla_x u| \to 0,$$

which means that for small values of $|\nabla_x u|$ the linear strain tensor $\varepsilon(\nabla_x u)$ is an approximation to the Green strain tensor. Replacing E by $\varepsilon(\nabla_x u)$ and neglecting the temperature dependence in (A.2.11) and (A.2.12) yields

$$\tilde{T} = \rho \nabla_\varepsilon \psi(\varepsilon(\nabla_x u), z) \tag{A.2.13}$$

$$\nabla_z \psi(\varepsilon(\nabla_x u), z) \cdot f(\varepsilon(\nabla_x u), z) \leq 0,$$

which are (2.1.8) and the dissipation inequality (2.1.9).

With this change (A.2.13) defines a new constitutive equation, which approximates the originally given constitutive equation (A.1.5) for small deformations. We do not discuss further simplifications and linearizations of this constitutive equation here, since such investigations are mainly of interest if one also takes into account the dependence of $\tilde{T}$ and thus of ψ on the internal variables z. The investigation of this dependence would lead to study the general theory of the formulation of constitutive equations modelling the inelastic behavior of solids. For this extended topic with ongoing research we refer the reader to [3, 6, 12, 13, 15, 50, 76, 82, 92, 93, 94, 95, 96, 97, 100, 113, 120, 132, 178, 180, 181, 208] and many other works in this field.

Bibliography

[1] H.-D. Alber: On a system of equations from the theory of nonlinear viscoplasticity. Preprint Nr. 1265, Fachbereich Mathematik, Technische Hochschule Darmstadt 1989

[2] H.-D. Alber: Global existence and boundedness of large solutions to nonlinear equations of viscoelasticity with hardening. Commun. Math. Phys. 166 (1995), 565–601

[3] H.-D. Alber: Mathematische Theorie des inelastischen Materialverhaltens von Metallen. Mitt. Ges. Angew. Math. Mech. 18 (1995), 9–38

[4] H.-D. Alber, M. Fuchs [eds.]: Workshop on the mathematical theory of nonlinear and inelastic material behavior, Technische Hochschule Darmstadt, Darmstadt, Germany, 1992. Bonn: Universität Bonn, Mathematisches Institut, Bonn. Math. Schr. 239 (1993)

[5] S.S. Antman: Nonlinear problems of elasticity. New York: Springer 1995

[6] S.S. Antman, G. W. Szymczak: Nonlinear elastoplastic waves. Contemp. Math. 100 (1989), 27–54

[7] P.J. Armstrong, C.O. Frederick: A mathematical representation of the multiaxial Bauschinger effect. C.E.G.B. Report RD/B/N 731, Berkeley Nuclear Laboratories 1966

[8] H.D. Baehr: Thermodynamik. Berlin: Springer 1988

[9] H. Bellout, F. Bloom, J. Nečas: Existence of global weak solutions to the dynamical problem for a three–dimensional elastic body with singular memory. SIAM J. Math. Anal. 24 (1993), 36–45

[10] A. Bensoussan, J. Frehse: Asymptotic behaviour of Norton-Hoff's law in plasticity theory and H_1 regularity. In: J.-L. Lions [ed.]: Boundary value problems for partial differential equations and applications. Dedicated to Enrico Magenes on the occasion of his 70th birthday. Paris: Masson, Res. Notes Appl. Math. 29 (1993), 3–25

[11] A. Bensoussan, J. Frehse: Asymptotic behaviour of the time dependent Norton-Hoff law in plasticity theory and H_1 regularity. Commentat. Math. Univ. Carol. 37 (1996), 285 – 304

[12] A. Bertram: Material systems – a framework for the description of material behavior. Arch. Rational Mech. Anal. 80 (1982), 99–133

[13] A. Bertram: Axiomatische Einführung in die Kontinuumsmechanik. Mannheim: BI Wissenschaftsverlag 1989

[14] A. Bertram, M. Kraska: Determination of finite plastic deformations in single crystals. Arch. Mech. 47 (1995), 203–222

[15] J.F. Besseling, E. van der Giessen: Mathematical modelling of inelastic deformation. London: Chapman and Hall 1994

[16] D. Blanchard, P. LeTallec: Numerical analysis of the equations of small strains quasistatic elastoviscoplasticity. Numer. Math. **50** (1986), 147–169

[17] D. Blanchard, P. LeTallec, M. Ravachol: Numerical analysis of evolution problems in nonlinear small strains elastoviscoplasticity. Numer. Math. **55** (1989), 177–195

[18] S.R. Bodner, Y. Partom: Constitutive equations for elastic–viscoplastic strain-hardening materials. J. Appl. Mech. **42** (1975), 385–389

[19] J.L. Bojarski: The relaxation of Signorini problems in Hencky plasticity, I: three dimensional solid. Nonlinear Anal., Theory Methods Appl. **29** (1997), 1091–1116

[20] J.L. Bojarski: The relaxation of Signorini problems in Hencky plasticity, II: plates. Nonlinear Anal., Theory Methods Appl. **29** (1997), 1117–1143

[21] H. Braasch: Ein Konzept zur Fortentwicklung und Anwendung viskoplastischer Werkstoffmodelle. Bericht 92-71, Institut für Statik der Technischen Universität Braunschweig. Braunschweig 1992

[22] H. Brézis: Multiplicateur de Lagrange en torsion élastoplastique. Arch. Rational Mech. Anal. **49** (1972), 32–40

[23] H. Brézis: Operateurs maximaux monotones. North Holland, Amsterdam, 1973

[24] M. Brokate, P. Krejčí: On the wellposedness of the Chaboche model. To appear in Proceedings of the 1996 Vorau conference on Control and Estimation of Distributed Parameter Systems. Basel: Birkhäuser

[25] M. Brokate, J. Sprekels: Hysteresis and phase transitions. New York: Springer 1996

[26] O.T. Bruhns, B. Boecke, F. Link: The constitutive relations of elastic-plastic materials at small strains. Nucl. Eng. and Design, **85** (1984), 325–331

[27] O.T. Bruhns, U. Rott: A viscoplastic model with a smooth transition to describe rate-independent plasticity. Int. J. Plast. **10** (1994), 347-362

[28] C. Carstensen: Coupling of FEM und BEM for interface problems in viscoplasticity and plasticity with hardening. SIAM J. Numer. Anal. **33** (1996), 171–207

[29] J.L. Chaboche: Viscoplastic constitutive equations for the description of cyclic and anisotropic behavior of metals. Bull. Acad. Polonaise Sci., Série Sci. Tech. **25** (1977), 33–39, 42–48

[30] J.L. Chaboche: Constitutive equations for cyclic plasticity and cyclic viscoplasticity. Int. J. Plast. **5** (1989), 247–302

[31] J.L. Chaboche: On some modifications of kinematic hardening to improve the description of ratchetting effects. Int. J. Plast. **7** (1991), 661–678

[32] J.L. Chaboche: Modeling of ratchetting: evaluation of various approaches. Eur. J. Mech. A/Solids **13** (1994), 501–518

[33] J.L. Chaboche, G. Rousselier: On the plastic and viscoplastic constitutive equations – Part I: Rules developed with internal variable concept. Trans. ASME J. Pressure Vessel Technol. **105** (1983), 153–158

[34] J.L. Chaboche, G. Rousselier: On the plastic and viscoplastic constitutive equations – Part II: Application of internal variable concepts to the 316 stainless steel. Trans. ASME J. Pressure Vessel Technol. **105** (1983), 159–164

[35] K. Chełmiński: Existence of large solutions to the quasistatic problem for a three-dimensional Maxwell material. Preprint Nr. 1577, Fachbereich Mathematik, Technische Hochschule Darmstadt 1993

[36] K. Chełmiński: Global in time existence of solutions of the constitutive model of Bodner-Partom with isotropic hardening. Demonstratio Math. **28** (1995), 667–688

[37] K. Chełmiński: On large solutions for the quasistatic problem in nonlinear viscoelasticity with the constitutive equations of Bodner-Partom. Math. Methods Appl. Sci. **19** (1996), 933-942

[38] K. Chełmiński: Energy estimates and global in time results for a problem from nonlinear viscoelasticity. Bull. Acad. Polonaise Sci., Série Math. **44** (1996), 465–477

[39] K. Chełmiński: Stress L^∞–estimates and the uniqueness problem for the quasistatic equations to the model of Bodner–Partom. Math. Methods Appl. Sci. **20** (1997), 1127 – 1134

[40] K. Chełmiński: Stress L^∞–estimates and the uniqueness problem for the equations to the model of Bodner–Partom in the two dimensional case. To appear in Math. Methods. Appl. Sci. (1997)

[41] K. Chełmiński: On initial-boundary value problems for the inelastic material behaviour of metals. To appear in Z. Angew. Math. Mech.

[42] K. Chełmiński, H.D. Alber: Existence theory for the equations of inelastic material behavior of metals – Transformation of interior variables and energy estimates. Rocz. Pol. Tow. Mat., Ser. III, Mat. Stosow **39** (1996), 1–15

[43] K. Chełmiński, P. Gwiazda: On the model of Bodner-Partom with nonhomogeneous boundary data. Submitted

[44] M. Chipot: Energy approximation. In: H.-D. Alber, M. Fuchs [eds.] (1993), 1–10

[45] S.H. Choi, E. Krempl: Viscoplastic theory based on overstress applied to the modeling of cubic single crystals. Eur. J. Mech., A/Solids **8** (1989), 219–233

[46] P.G. Ciarlet: Lectures on three-dimensional elasticity. New-York: Springer 1983

[47] P.G. Ciarlet: Mathematical elasticity. Volume I: Three dimensional elasticity. Amsterdam: North–Holland 1988

[48] B.D. Coleman, M.E. Gurtin: Thermodynamics with internal state variables. J. Chem. Phys. **47** (1967), 597–613

[49] D. Cordts, F.G. Kollmann: An implicit time integration scheme for inelastic constitutive equations with internal state variables. Internat. J. Numer. Methods Engrg. **23** (1986), 533–554

[50] N. Cristescu, I. Suliciu: Viscoplasticity. The Hague: Martinus Nijhoff Publishers 1982

[51] C. Dafermos: Global smooth solutions to the initial-boundary value problem for the equations of one–dimensional nonlinear thermoviscoelasticity. SIAM. J. Math. Anal. **13** (1982), 397–408

[52] C. Dafermos: Dissipation in materials with memory. In: A.S. Lodge, M. Renardy, J.A. Nohel (eds.), Viscoelasticity and rheology, Proc. Symp., Madison/Wis. 1984, Publ. Math. Res. Cent. Univ. Wis. Madison **53** (1985), 221–234

[53] C. Dafermos: Development of singularities in the motion of materials with fading memory. Arch. Rational Mech. Anal. **91** (1986), 193–205

[54] C. Dafermos: Hyperbolic conservation laws with memory. Differential equations, Proc. EQUADIFF Conf., Xanthi/Greece 1987, Lect. Notes Pure Appl. Math. **118** (1989), 157–166

[55] C. Dafermos, L. Hsiao: Development of singularities in solutions of the equations of nonlinear thermoelasticity. Quart. Appl. Math. **44** (1986), 463–474

[56] C. Dafermos, J.A. Nohel: Energy methods for nonlinear hyperbolic volterra integrodifferential equations. Comm. Partial Differential Equations **4** (1979), 219–278

[57] C. Dafermos, J.A. Nohel: A nonlinear hyperbolic volterra equation in viscoelasticity. Contributions to analysis and geometry, Suppl. Amer. J. Math. (1981), 87–116

[58] G. Dinca: Sur la monotonie d'après Minty-Browder de l'opérateur de la théorie de la plasticité. C.R. Acad. Sci. Paris, Sér. A, **269** (1969), 535–538

[59] G. Dinca: Opérateurs monotones dans la théorie de plasticité. Ann. Scuola Norm. Sup. Pisa, Sci. Fis. Mat. **24** (1970), 357–399

[60] D.C. Drucker: A more fundamental approach to plastic stress strain relations. In: Proc. 1st. U.S. Natl. Congress Appl. Mech. p. 478–491, 1951

[61] N. Dunford, J.T. Schwartz: Linear operators Part I: General theory. New York: Wiley-Interscience 1988

[62] G. Duvaut, J. L. Lions: Les inéquations en méchanique et en physique. Paris: Dunod 1972

[63] J. Eftis, M. S. Abdel-Kader, D. L. Jones: Comparisons between the modified Chaboche and Bodner-Partom viscoplastic constitutive theories at high temperature. Int. J. Plast. **5** (1989), 1–27

[64] Y. Estrin, H. Mecking: A unified phenomenological description of work hardening and creep based on one-parameter models. Acta Metall. **32** (1984), 57–70

[65] L.C. Evans: Nonlinear evolution equations in an arbitrary Banach space. Israel J. Math. **26** (1977), 1–42

[66] R.A. Eve, B.D. Reddy, R.T. Rockafellar: An internal variable theory of elastoplasticity based on the maximum plastic work inequality. Quarterly Appl. Math. **48** (1990), 59–83

[67] S. Flügge [ed.]: Handbuch der Physik. Berlin: Springer 1956 ff.

[68] A.D. Freed, J.L. Chaboche: A viscoplastic theory with thermodynamic considerations. Acta Mech. **90** (1991), 155-174

[69] A. Friedman: Variational principles and free boundary value problems. New York: Wiley 1982

[70] M. Fuchs: On quasi-static non Newtonian fluids with power-law. Math. Methods Appl. Sci. **19** (1996), 1225–1232

[71] M. Fuchs, J.F. Grotowski, J. Reuling: On variational models for quasi-static Bingham fluids. Math. Methods Appl. Sci. **19** (1996), 991-1015

[72] M. Fuchs, G. Seregin: Partial regularity of the deformation gradient for some model problems in nonlinear two-dimensional elasticity. St. Petersbg. Math. J. **6** (1995), 1229-1248

[73] S. Fučik, A. Kufner [eds.]: Nonlinear analysis. Function spaces and applications. Leipzig: Teubner 1979

[74] H. Gajewski: Über einige Fehlerabschätzungen bei Gleichungen mit monotonen Potentialoperatoren in Banachräumen. Monatsberichte Akad. d. Wiss. Berlin (1970), 571–579

[75] H. Geiger, K. Scheel [eds.]: Handbuch der Physik, Vols. 1–24. Berlin: Springer 1926 ff.

[76] H. Geiringer: Ideal plasticity. In: S. Flügge [ed.] Vol. VIa/3, 403–533, 1972

[77] C. Gerhard: On the existence and uniqueness of a warpening function in the elasto-plastic torsion of a cylindrical bar with multiply connected cross section. In: P. Germain, B. Nayroles [eds.] 328–342, 1976

[78] P. Germain, B. Nayroles [eds.]: Applications of functional analysis to problems in mechanics. Lecture Notes in Math. **503**. Berlin, Heidelberg, New York: Springer 1976

[79] P. Germain, Q.S. Nguyen, P. Suquet: Continuum thermodynamics. J. Appl. Mech. **50** (1983), 1010–1019

[80] R. Glowinski: Lectures on numerical methods for nonlinear variational problems. New York: Springer 1980

[81] R. Glowinski: Numerical methods for nonlinear variational problems. New York: Springer 1984

[82] A.E. Green, P.M. Naghdi: A general theory of an elastic–plastic continuum. Arch. Rational Mech. Anal. **18** (1965), 251–281

[83] J.M. Greenberg: Models of elastic–perfectly plastic materials. Eur. J. Appl. Math. **1** (1990), 131–150

[84] K. Gröger: Zur Theorie des quasi-statischen Verhaltens von elastisch-plastischen Körpern. Z. Angew. Math. Mech. **58** (1978), 81–88

[85] K. Gröger: Initial-value problems for elastoplastic and elasto-viscoplastic systems. In: S. Fučik and A. Kufner [eds.], 95–127, 1979

[86] M. Gurtin: The linear theory of elasticity. In: S. Flügge, [ed.]: Vol VIa/2, 1–296, 1972

[87] A. Haar, T. v. Kármán: Theorie der Spannungszustände in plastischen und sandartigen Medien. Göttinger Nachrichten (1909), 204–218

[88] B. Halphen, Nguyen Quoc Son: Sur les matériaux standards généralisés. J. Méc. **14** (1975), 39–63

[89] E.W. Hart: A phenomenological theory for plastic deformation of polycrystalline metals. Acta Metall. **18** (1970), 599–610

[90] E.W. Hart: Constitutive relations for the nonelastic deformation of metals. Trans. ASME J. Engrg. Mat. Technol. **98** (1976), 193–202

[91] G. Hartmann, F. G. Kollmann: A computational comparison of the inelastic constitutive models of Hart and Miller. Acta Mech. **69** (1987), 139–165

[92] P. Haupt: Viskoelastizität und Plastizität. Berlin: Springer 1977

[93] P. Haupt: On the concept of an intermediate configuration and its applications to a representation of viscoelastic–plastic material behavior. Int. J. Plast. **1** (1985), 303–316

[94] P. Haupt: Foundation of continuum mechanics. In: K. Hutter [ed.], 1–77, 1993

[95] P. Haupt: Thermodynamics of Solids. In: W. Muschik [ed.], 65–137, 1993

[96] P. Haupt: On the mathematical modelling of material behavior in continuum mechanics. Acta Mech. **100** (1993), 129–154

[97] P. Haupt, M. Kamlah, Ch. Tsakmakis: Continuous representation of hardening properties in cyclic plasticity. Int. J. Plast. **8** (1992), 803–817

[98] H. Hencky: Zur Theorie plastischer Deformationen. Z. Angew. Math. Mech. **4** (1924), 323–334

[99] G.A. Henshall, A.K. Miller: Simplifications and improvements in unified constitutive equations for creep and plasticty. Part 1. Equations development. Acta Metall. Mater. **38** (1990), 2101–2115

[100] R. Hill: The mathematical theory of plasticity. Oxford: Clarendon Press 1950

[101] W.J. Hrusa, J.A. Nohel: The Cauchy problem in one-dimensional nonlinear viscoelasticity. J. Differ. Equations **59** (1985), 388–412

[102] R. Hünlich: On simultaneous torsion and tension of a circular cylindrical bar consisting of an elastoplastic material with linear hardening. Z. Angew. Math. Mech. **58** (1979), 509–516

[103] K. Hutter: The foundation of thermodynamics, its basic postulates and implications. A review of modern thermodynamics. Acta Mech. **27** (1977), 1–54

[104] K. Hutter [ed.]: Continuum mechanics in environmental sciences and geophysics. CISM Courses and Lectures **337**. Wien: Springer 1993

[105] K. Hutter: Fluid- und Thermodynamik. Berlin. Springer 1995

[106] I. R. Ionescu, M. Sofonea: Functional and numerical methods in viscoplasticity. Oxford: Oxford University Press, 1993

[107] C. Johnson: Existence theorems for plasticity problems. J. Math. Pures Appl. **55** (1976), 431–444

[108] C. Johnson: On plasticity with hardening. J. Math. Anal. Appl. **62** (1978), 325–336

[109] T. Kato: Perturbation theory for linear operators. New York: Springer 1966

[110] F. Klaus: Lokaler Existenz- und Eindeutigkeitssatz für die Millerschen Gleichungen zur nichtlinearen Viskoelastizität mit Härtung. Dissertation, Fachbereich Mathematik, Technische Hochschule Darmstadt, 1994

[111] F. Klaus: Viscoelasticity with hardening. To appear in Math. Methods Appl. Sci. (1997)

[112] U.F. Kocks: Laws for work-hardening and low-temperature creep. Trans. ASME J. Engrg. Mat. Technol. **98** (1976), 76–85

[113] W.T. Koiter: General theorems for elastic-plastic solids. In: I.N. Sneddon, R. Hill [eds.], 165–220, 1960

[114] A. Korn: Sur les équations d'élasticité. Ann. École Norm. **24** (1907), 9–75

[115] V. Korneev, U. Lange: Approximate solution of plastic flow theory problems. Leipzig: Teubner 1984

[116] M. Korzen: Beschreibung des inelastischen Materialverhaltens im Rahmen der Kontinuumsmechanik: Vorschlag einer Materialgleichung vom viskoelastisch-plastischen Typ. Dissertation, Fachbereich Mechanik, Technische Hochschule Darmstadt, 1988

[117] G. Kracht: Erschließung viskoplastischer Stoffmodelle für thermomechanische Strukturanalysen. Bericht 93-69, Institut für Statik der Technischen Universität Braunschweig. Braunschweig 1993

[118] M. Krasnoselskii, A. Pokrovskii: Systems with hysteresis. Moscow: Nauka 1983

[119] J. Kratochvil, O.W. Dillon: Thermodynamics of elastic-plastic materials as a theory with internal state variables. J. Appl. Phys. **40** (1969), 3207–3218

[120] A. Krawietz: Materialtheorie. Berlin: Springer 1986

[121] P. Krejčí: Hysteresis and periodic solutions of semilinear and quasilinear wave equations. Math. Z. **193** (1986), 247–264

[122] P. Krejčí: Hysteresis memory preserving operators. Apl. Mat. **36** (1991), 305–326

[123] P. Krejčí: Hysteresis, convexity and dissipation in hyperbolic equations. Tokyo: Gakkotosho 1996

[124] E. Krempl, J.J. McMahon, D. Yao: Viscoplasticity based on overstress with a differential growth law for the equilibrium stress. Mech. Mater. **5** (1986), 35–48

[125] E. Kröner: Mikrostrukturmechanik. Mitt. Ges. Angew. Math. Mech. **15** (1992), 104 110

[126] M.S. Kuczma, E. Stein: On nonconvex problems in the theory of plasticity. Arch. Mech. (Arch. Mech. Stos.) **46** (1994), 505–529

[127] P. Laborde: On visco-plasticity with hardening. Numer. Funct. Anal. Optimi. 1 (1979), 315–339

[128] H. Lanchon: Torsion élastoplastique d'une barre cylindrique de section simplement ou multiplement connexe. J. Mech. **13** (1974), 267–320

[129] A. Langenbach: Monotone Potentialoperatoren. Berlin: Verlag der Wissenschaften 1976

[130] T. Lehmann: On thermodynamically-consistent constitutive laws in plasticity and viscoplasticity. Arch. Mech. (Arch. Mech. Stos.) **40** (1988), 415–431

[131] R. Leis: Initial boundary value problems in mathematical physics. Chichester, Stuttgart: Wiley – Teubner 1986

[132] J. Lemaitre, J.L. Chaboche: Mechanics of solid materials. Cambridge: Cambridge University Press 1994. (French edition: Paris: Dunod 1985)

[133] P. LeTallec: Numerical analysis of viscoelastic problems. Paris: Masson; Berlin: Springer 1990

[134] M. Lévy: Mémoires sur les équations des corps solides ductiles au-déla de la limite élastique. J. Math. Pures Appl. **16** (1871), 369–372

[135] L. Lichtenstein: Über die erste Randwertaufgabe der Elastizitätstheorie. Math. Z. **20** (1924), 21–28

[136] I-Shih Liu: Method of Lagrange multipliers for exploitation of the entropy principle. Arch. Rational Mech. Anal. **46** (1972), 131–148

[137] I-Shih Liu: On entropy flux-heat flux relation in thermodynamics with Lagrange multipliers. Contin. Mech. Thermodyn. **8** (1996), 247-256

[138] M.C.M. Liu, E. Krempl: A uniaxial viscoplastic model based on total strain and overstress. J. Mech. Phys. Solids **27** (1979), 377–391

[139] I-Shih Liu, I. Müller: Thermodynamics of mixtures of fluids. In: C. Truesdell [ed.], 264–286, 1984

[140] J. Lubliner: On fading memory in materials of evolutionary type. Acta Mech. **8** (1969), 75–81

[141] R.C. MacCamy: A model for one-dimensional, nonlinear viscoelasticity. Quart. Appl. Math. **35** (1977), 21–33

[142] T. Malmberg: Thermodynamics of a visco-plastic model with internal variables. KfK 4572, Kernforschungszentrum Karlsruhe 1990

[143] J. Marsden, T. Hughes: Mathematical foundations of elasticity. Englewood Cliffs: Prentice-Hall 1983

[144] L. Méric, G. Cailletaud: Single crystal modeling for structural calculations: Part 2 – finite element implementation. Trans. ASME J. Engrg. Mat. Technol. **113** (1991), 171–182

[145] L. Méric, P. Poubanne, G. Cailletaud: Single crystal modeling for structural calculations: Part 1 – model presentation. Trans. ASME J. Engrg. Mat. Technol. **113** (1991), 162–170

[146] E. Miersemann: Zur Regularität der quasistatischen elasto-viskoplastischen Verschiebungen und Spannungen. Math. Nachr. **96** (1980), 293–299

[147] A.K. Miller: An inelastic constitutive model for monotonic, cyclic and creep deformation: Part I – Equations development and analytical procedures. Trans. ASME J. Engrg. Mat. Technol. **98** (1976), 97–105

[148] A.K. Miller: An inelastic constitutive model for monotonic, cyclic, and creep deformation: Part II – Application to type 304 stainless steel. Trans. ASME J. Engrg. Mat. Technol. **98** (1976), 106–113

[149] A.K. Miller [ed.]: Unified constitutive equations for plastic deformation and creep of engineering alloys. New York: Elsevier 1987

[150] R. v. Mises: Mechanik der festen Körper im plastisch–deformablen Zustand. Nachr. Akad. Wiss. Göttingen Math.-Phys. Kl. (1913), 582–592. (Selected papers of Richard von Mises I. Providence: American Mathematical Society 1963, 189–199)

[151] R. v. Mises: Mechanik der plastischen Formänderung von Kristallen. Z. Angew. Math. Mech. **8** (1928), 161–185

[152] T. Miyoshi: Foundations of the numerical analysis of plasticity. Amsterdam: North-Holland 1985

[153] J.J. Moreau: La notion de sur-potentiel et les liaisons unilatérales en élastostatique. C. R. Acad. Sci. Paris Sér. A, **267** (1968), 954–957

[154] J.J. Moreau: Sur l'évolution d'un système élasto-visco-plastique. C. R. Acad. Sci. Paris, Sér. A, **273** (1971), 118–121

[155] J.J. Moreau: On unilateral constraints, friction and plasticity. In: G. Capriz, G. Stampacchia [eds.]: New variational techniques in mathematical physics. Centro Internationale Matematico Estivo, II Ciclo 1973, Roma: Edizioni Cremonese 1974, 175–322

[156] J.J. Moreau: Application of convex analysis to the treatment of elastoplastic systems. In: P. Germain, B. Nayroles [eds.] 56–89, 1976

[157] J.J. Moreau: Evolution problem associated with a moving convex set in a Hilbert space. J. Differ. Equations **26** (1977), 347–374

[158] Z. Mroz et al.: A non-linear hardening model and its application to cyclic loading. Acta Mech. **25** (1976), 51–61

[159] I. Müller: Thermodynamics. Boston: Pitman 1985

[160] S. Müller, S.J. Spector: Existence of singular minimizers in three-dimensional non-linear elasticity In: H.-D. Alber, M. Fuchs [eds.] 21–31, 1993

[161] W. Muschik [ed.]: Non-equilibrium with applications to solids. CISM Courses and Lectures **336**. Wien: Springer 1993

[162] B. Nayroles: Essai de théorie fonctionnelle des structures rigides plastiques parfaites. J. Méc. **9** (1970), 491–506

[163] B. Nayroles: Quelques applications variationnelles de la théorie des fonctions duales à la Mécanique des solides. J. Méc. **10** (1971), 263–289

[164] J. Nečas, I. Hlaváček: Mathematical theory of elastic and elastico–plastic bodies: An introduction. Amsterdam: Elsevier 1981

[165] Nguyen Quoc Son: Matériau elastoplastique ecrouissable. Distribution de la contrainte dans une evolution quasi-statique. Arch. Mech. (Arch. Mech. Stos.) **25** (1973), 695–702

[166] Nguyen Quoc Son: On the elastic plastic initial-boundary value problem and its numerical integration. Internat. J. Numer. Methods. Engrg. **11** (1977), 817–832

[167] Nguyen Quoc Son, D. Radenkovic: Stability of equilibrium in elastic-plastic solids. In: P. Germain, B. Nayroles [eds.] 403–414, 1976

[168] W.A. Noll: A mathematical theory for the mechanical behavior of continuous media. Arch. Rational Mech. Anal. **2** (1958/59), 197–226

[169] D. Nouailhas: A viscoplastic modeling applied to stainless steel behavior. Proceedings of the second international conference on constitutive laws for engineering materials: Theory and applications, Tucson, Arizona, U.S.A. 1987

[170] D. Nouailhas, A.D. Freed: A viscoplastic theory for anisotropic materials. Trans. ASME J. Engrg. Mat. Technol. **114** (1992), 97–104

[171] A. Nouri, M. Rascle: A global existence and uniqueness theorem for a model problem in dynamic elasto–plasticity with isotropic strain-hardening. SIAM J. Math. Anal. **26** (1995), 850–868.

[172] O.A. Oleinik, A.S. Shamaev, G.A. Yosifian: Mathematical problems in elasticity and homogenization. Amsterdam: North Holland 1992

[173] E.T. Onat, D.C. Drucker: Inelastic instability and incremental theories of plasticity. J. Aeronaut. Sci. **20** (1953), 181–186

[174] M. Ortiz, J.C. Simo: An analysis of a new class of integration algorithms for elasto-plastic constitutive relations. Internat. J. Numer. Methods Engrg. **23** (1986), 353–366

[175] P. Perzyna: On the constitutive equations for work-hardening and rate sensitive plastic materials. Bull. Acad. Polonaise Sci., Série. Sci. Tech. **12** (1964), 199–206

[176] P. Perzyna: Fundamental problems in viscoplasticity. Adv. Appl. Mech. **9** (1966), 243–377

[177] A.R.S. Ponter, F.A. Leckie: Constitutive relations for the time-dependent deformation of metals. Trans. ASME J. Engrg. Mat. Technol. **98** (1976), 47–51

[178] W. Prager: Probleme der Plastizitätstheorie. Basel: Birkhäuser 1955 (English edition: Problems in plasticity, London: Addison-Wesley 1959)

[179] W. Prager: A new method of analyzing stresses and strains in work-hardening plastic solids. J. Appl. Mech. **23** (1956), 493–496

[180] W. Prager: Einführung in die Kontinuumsmechanik. Basel: Birkhäuser 1961

[181] W. Prager, P. Hodge: Theory of perfectly plastic solids. New York: Wiley 1951

[182] L. Prandtl: Anwendungsbeispiel zu einem Henckyschen Satz über das plastische Gleichgewicht. Z. Angew. Math. Mech. **3** (1923), 401-406. (Gesammelte Abhandlungen. Berlin: Springer 1961, 113-121)

[183] L. Prandtl: Spannungsverteilung in plastischen Körpern. Proc. Int. Congr. Appl. Mech. Delft 1924. (1925), 43-54 (Gesammelte Abhandlungen. Berlin: Springer 1961, 133-148)

[184] G. Propst, J. Prüß: On wave equations with boundary dissipation of memory type. J. Integral Equations Appl. **8** (1996), 99-123

[185] J. Prüß: Evolutionary integral equations and applications. Basel: Birkhäuser 1993

[186] J. Prüß: Stability of linear hyperbolic viscoelasticity. In: H.-D. Alber, M. Fuchs [eds.] 43-52, 1993

[187] R. Racke: Lectures on nonlinear evolution equations. Braunschweig: Vieweg 1992

[188] R. Racke, Songmu Zheng: Global existence and asymptotic behavior in nonlinear thermoviscoelasticity. J. Differ. Equations **134** (1997), 46-67

[189] M. Rascle: Global existence of L^2-solutions in dynamical elasto-plasticity. Mat. Contemp. **11** (1996), 121-134.

[190] M. Renardy, W.J. Hrusa, J.A. Nohel: Mathematical problems in viscoelasticity. Harlow: Longman 1987

[191] A. Reuss: Berücksichtigung der elastischen Formänderung in der Plastizitätstheorie. Z. Angew. Math. Mech. **10** (1930), 266-274

[192] D.N. Robinson: A unified creep-plasticity model for structural metals at high temperature. ORNL, TM-5969, 1978

[193] D.N. Robinson, P.A. Bartolotta: Viscoplastic constitutive relationships with dependence on thermomechanical history. NASA CR-174836 (1985)

[194] M. Růžička: Multipolar materials. In: H.-D. Alber, M. Fuchs [eds.] 53-64, 1993

[195] M. de Saint-Venant: Sur les équations du mouvement intérieur des solides ductiles. J. Math. Pures Appl. **16** (1871), 373-382

[196] I.N. Sneddon, R. Hill [eds.]: Progress in Solid Mechanics, Vol. 1. Amsterdam: North-Holland 1960

[197] E. Steck: A stochastic model for high temperature plasticity of metals. Int. J. Plast. **1** (1985), 243-258

[198] E. Stein, Y. Huang: An analytical method to solve shakedown problems with linear kinematic hardening materials. Int. J. Solids Structures, **18** (1994), 2433-2444

[199] E. Stein, P. Wriggers: "Computational Mechanics" bei Festkörpern und Ingenieurstrukturen unter Verwendung von Finite-Element-Methoden. Mitt. Ges. Angew. Math. Mech. **12** (1989), 3-39

[200] D.C. Stouffer, S.R. Bodner: A constitutive model for the deformation induced anisotropic plastic flow of metals. Int. J. Engrg. Sci. **17** (1979), 737

[201] P.-M. Suquet: Evolution problems for a class of dissipative materials. Quart. Appl. Math. **38** (1980/81), 391-414

[202] P.-M. Suquet: Sur les équations de la plasticité: Existence et régularité de la solution. J. Méc. **20** (1981), 1–39

[203] M. Sütçü, E. Krempl: A stability analysis of the uniaxial viscoplasticity theory based on overstress. Computational Mech. **4** (1989), 401–408

[204] J.C. Swearengen, J.H. Holbrook: Internal variable models for rate-dependent plasticity: Analysis of theory and experiments. Res. Mech. **13** (1985), 93–128

[205] R. Temam: Problèmes mathématiques en plasticité. Paris: Gauthier–Villars 1983

[206] R. Temam: A generalized Norton-Hoff model and the Prandtl-Reuss law of plasticity. Arch. Rational Mech. Anal. **95** (1986), 137–183

[207] T. Ting: Elastic-plastic torsion. Arch. Rational Mech. Anal. **34** (1969), 228–244

[208] T. Ting: Topics in mathematical theory of plasticity. In: S. Flügge [ed.], Vol. VIa/3, 535–623, 1972

[209] E. Trefftz: Mathematische Elastizitätstheorie. In: H. Geiger, K. Scheel [eds.] Vol. 6, 47–140, 1926

[210] H. Tresca: C.R. Acad. Sci. Paris **59** (1864), 754

[211] C. Truesdell [ed.]: Rational thermodynamics. 2nd ed. New York: Springer 1984

[212] A. Visintin: Differential models of hysteresis. Berlin: Springer 1994

[213] A. Visintin: Models of phase transition. Boston: Birkhäuser 1996

[214] D. Yao, E. Krempl: Viscoplastic theory based on overstress. The prediction of monotonic and cyclic proportional and nonproportional loading paths of an aluminium alloy. Int. J. Plast. **1** (1985), 259–274

[215] E. Zeidler: Nonlinear functional analysis and its applications IV. (Applications to mathematical physics.) New York: Springer 1988

Index

Printing: Druckhaus Beltz, Hemsbach
Binding: Buchbinderei Schäffer, Grünstadt